Advances in Bioelectrochemistry Volume 3

Frank Nelson Crespilho
Editor

Advances in Bioelectrochemistry Volume 3

Biosensors, Wearable Devices and Biomedical Applications

Springer

Editor
Frank Nelson Crespilho
University of Sao Paulo
São Carlos, São Paulo, Brazil

ISBN 978-3-030-97920-1 ISBN 978-3-030-97921-8 (eBook)
https://doi.org/10.1007/978-3-030-97921-8

This Springer imprint is published by the registered company Springer Nature Switzerland AG
The registered company address is: Gewerbestrasse 11, 6330 Cham, Switzerland

Contents

Field-Effect Transistors for Biomedical Applications

Edson Giuliani Ramos Fernandes, Henrique Antonio Mendonça Faria, and Nirton Cristi Silva Vieira

Abstract Field-effect transistors are devices suitable for applications in bioelectronics, because of their small size, high signal-to-noise ratio, and the possibility of using biocompatible and flexible materials in device design. Since the introduction of the ion-sensitive field-effect transistor (ISFET) by Bergveld, various types of FETs are applied as transducers for the development of biomedical devices. These include the ISFET and its derivatives, such as the extended gate field-effect transistors (EGFETs), the organic field-effect transistors (OFETs), nanomaterial-based field-effect transistors (NanoFETs), and the tunnel field-effect transistors (TFETs). In this chapter, we highlight FET devices for biomedical applications. A description is made of the application of FETs directly with biological systems, including biosensors, interfaces with cells and tissues, and other biomedical applications. We also provide challenges and perspectives for this fascinating technology.

Keywords Field-effect transistor · Biosensor · Transduction · Bioelectronics

1 Introduction

Bioelectronics is a research field that has a remarkable multidisciplinary feature. In general, bioelectronics combines biological materials and electronic components in the search for healthcare solutions, including diagnostic and therapeutic methods for diseases and the monitoring of biological systems [1]. Good examples of the development of bioelectronics are biomedical devices such as glucose meter sensors, brain stimulators, and cardiac pacemakers. The synergy of biological systems with electronic transducers is essential for the development of new biomedical devices.

E. G. R. Fernandes · N. C. S. Vieira (✉)
Institute of Science and Technology, Federal University of São Paulo, São José dos Campos, SP 12231-280, Brazil
e-mail: ncsvieira@unifesp.br

H. A. M. Faria
Institute of Chemistry, São Paulo State University (Unesp), Araraquara, SP 14800-060, Brazil

F. N. Crespilho (ed.), *Advances in Bioelectrochemistry Volume 3*,
https://doi.org/10.1007/978-3-030-97921-8_1

Reducing the dimensions of electronic components for this purpose has been a challenge. Still, it seems to be the way forward in the search for improvements in terms of sensitivity, biocompatibility, and extension of the applicability of these devices [2].

Field-effect transistors are devices suitable for applications in bioelectronics, because of their small size, high signal-to-noise ratio, and the possibility of using biocompatible and flexible materials in device design. Since the introduction of the ion-sensitive field-effect transistor (ISFET) by Bergveld, various types of FETs are applied as transducers for the development of biomedical devices. These include the ISFET and its derivatives, such as the extended gate field-effect transistor (EGFETs), the organic field-effect transistors (OFETs), nanomaterial-based field-effect transistors (NanoFETs), and the tunnel field-effect transistors (TFETs). In this chapter, we highlight FET devices for biomedical applications. A description is made of the application of FETs directly with biological systems, including biosensors, interfaces with cells and tissues, and other biomedical applications. We also provide challenges and perspectives for this fascinating technology.

2 Biomedical Applications

2.1 Biosensors

Biosensors are analytical devices that incorporate selective biological elements (capable of recognizing target analytes specifically) in intimal contact to a transducer and convert the biological response into an electrical signal. Figure 1 shows the incorporation of biological molecules using graphene as an example of a biosensor platform. The biosensor idea was first published by Clark and Lyons, in [3], proposing the entrapping glucose oxidase in a dialysis membrane over an electrochemical electrode [3]. In 1969, Guilbault and Montalvo reported the fabrication of a urea biosensor based on ion-selective electrodes using glass electrodes coupled with urease [4].

The first biosensor using an ISFET as a transducer was proposed in 1976 by Janata and Moss [6] and realized four years later, in 1980, by Caras and Janata for penicillin sensing [7]. After that, the immobilization of the biological element on the support of reduced dimensions was the critical point for the realization of FET biosensors since, as in any biosensor, the biological activity must be maintained, and non-specific adsorption avoided. According to the biological recognition material, the following sections deal with the progress made for the most different types of FET biosensors: FETs modified with enzymes (EnFETs), with antibodies and antigen (ImmunoFETs), with nucleic acid (DNA-FETs), and an overview of the recent CRISPR (clustered regularly interspaced short palindromic repeats) and CRISPR-associated proteins (Cas) technology.

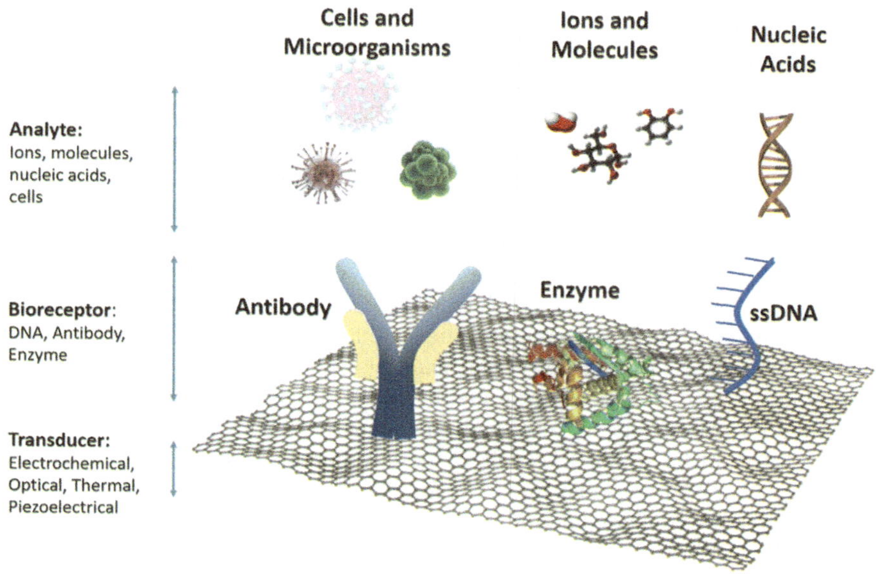

Fig. 1 Examples of biosensors and components on a graphene platform. Reprinted with permission from reference [5]. Copyright (2018) Springer Nature

2.1.1 EnFETs

Enzymes are catalytic biomolecules, and the interest to incorporate them in FET devices is their selectivity and efficient catalysis for the target analyte—particularly in a crude biological sample. In essence, the enzyme-modified FETs (EnFETs) are based on the enzyme immobilization near the gate insulator of a FET capable of producing or consuming protons during catalysis and, then, response to the local pH variations [8]. As reported in the last section, the first EnFET was proposed by Janata and Moss [6] and later developed by Caras and Janata for penicillin sensing [7]. The working principle of this EnFET is based on the biocatalytic reaction of the penicillinase enzyme with its substrate, penicillin, generating or consuming reagents that can modulate the drain current. The enzyme penicillinase hydrolyzes penicillin into penicilloic acid, changing the local pH as follows:

$$\text{penicillin} + H_2O \xrightarrow{\text{penicillinase}} \text{penicillinoic acid} + H^+$$

The production of protons on the ISFET gate decreases the local pH resulting in an increased drain current of the EnFET [7].

The most studied EnFETs are based on glucose oxidase, urease, and penicillinase because of the historical importance of detecting the respective substrates: glucose, urea, and penicillin in the biomedical area. These types of sensors become suitable model systems for other EnFETs. Then, the detection principle can be expanded for other enzymes (and respective substrates), e.g., glucose oxidase (glucose), glucose dehydrogenase (glucose, sucrose, maltose), β-galactosidase (lactose), invertase (sucrose), maltase (maltose), urease (urea), alcohol dehydrogenase (ethanol), alcohol oxidase (formaldehyde), peroxidase (ascorbic acid), organophosphate hydrolase (organophosphate compound), and acetylcholinesterase (acetylcholine) [9].

Once the enzyme immobilization is the core of a biosensor, several methods have been reported in the literature to immobilize enzymes on the gate surface of an ISFET, on the extended gate of EGFET/SEGFET, on OFET or NanoFETs. These can be classified as physical methods (reversible) by entrapment, adsorption, and encapsulation, or chemical processes (irreversible) by covalent bonding and crosslinking [9]. Besides sensitivity, the immobilization method also determines the reuse of the EnFET and their both response time and stability. Each method has advantages and disadvantages [8]. For example, in physical methods, the benefits are that the enzyme does not chemically react with the support (which could compromise its active site), and the simplicity, such as the adsorption technique, avoids the use of harmful chemical reagents. However, the possible leakage of enzymes by weak immobilization can be critical and reduce the sensitivity and reproducibility of the EnFET [9]. Covalent immobilization of enzymes involves an effective binding to the support through prior activation of functional groups allowing a good sensitivity of the EnFET. However, using some bonding agents can make cross-linking between enzymes possible, leading to loss of efficiency by blocking the active site or inactivating the enzymes [10].

The current trends in EnFETs fabrication are maintaining sensor stability, sensitivity (nM to aM), reproducibility, sufficient lifetime, and avoiding the leakage of the enzyme layer near the gate surface. To overcome drawbacks relative to enzymes activity and stability, enzymeless biosensors based on biological mimicry systems were proposed. For example, Huang et al. reported a flexible and reusable NanoFET (over 10 times) based on graphene for glucose detection with a limit of detection (LOD) of 0.15 mM and a dynamic range of 0.05–100 mM. In the paper, the authors used pyrene-1-boronic acid (PBA) as the glucose receptor aiming for wearable sensing applications [11]. Park and colleagues reported a FET based on graphene–polypyrrole nanotube composites as the conductive channel for H_2O_2 detection [12].

Minami et al. proposed an EGFET using an OFET device to detect biogenic amines (histamine and putrescine) in aqueous solution [13]. The organic EGFET was based on the diamine oxidase, and horseradish peroxidase was immobilized using an osmium-redox polymer as a mediator. The extended gate electrode was fabricated by thermal evaporation of gold (50 nm) on a film substrate of polyethylene naphthalate and modified with deposition of horseradish peroxidase osmium-redox polymer and diamine oxidase. The biosensor response was due to the I_{DS} changes in the presence of histamine with a linear range up to 10 μM and a limit of detection of 1.2 μM [13].

This section discussed the basic principles of EnFETs with an emphasis on enzyme immobilization, efficiency, and applications. Next, the antibodies and antigens in FET platforms will be explored.

2.1.2 ImmunoFETs

Antibodies belong to a family of glycoproteins known as immunoglobulins (Ig), separated into five classes: IgA, IgG, IgM, IgD, and IgE. IgG is the amplest in the body, corresponding to a fraction of 70% of the antibodies. IgG is a Y-shaped molecule made up of two polypeptide chains linked together by disulfide bonds. There are two identical chains named heavy chains (~50 kDa each), and two identical chains called light chains (~25 kDa each) totaling about 150 kDa [14]. Polyclonal antibodies originate from immune cells. Therefore, they are a mixture of immunoglobulins against a specific antigen, with each one recognizing an epitope, they react with several epitopes of the same antigen [15]. On the other hand, monoclonal antibodies are produced from immune cells equal to parental ones and recognize specific epitopes of an antigen and, for this reason, are more selective than polyclonal antibodies [15].

Immunosensors are biosensors that use immobilized antibodies or antigens to recognize one of these molecules. Then, the transducer converts the antibody–antigen reaction into a measurable signal. FET transducers are ideal to be applied as immunosensors because they can monitor antibody–antigen interactions directly. Without the label, such enzymes or chromophores are linked to secondary antibodies to generate a detectable signal in sandwich-like immunosensors. ImmunoFETs refer to FETs modified with antibodies or antigens on the gate oxide in the case of the ISFET or directly on the nanomaterial in the case of the NanoFETs [16].

For the specific case of detection of antigen–antibody interactions in FETs, the general interpretation is valid: Antigens and antibodies are electrically charged organic molecules. The polarity and intensity of the charge depend on the pH of the solution. When an antibody (Ab) molecule charged with charge C1 binds to the antigen (Ag) molecule charged with charge C2, the formed Ab-Ag complex contains a new charge C3 [17] (which is not necessarily the sum of both). This charge change could, in principle, be detected by a FET. The immobilization of one of these biomolecules, antibodies or antigens, on FET, will enable the recognition of the other (and vice versa), and the immunoreaction is responsible for the alteration of the potential on the surface of the FET and the consequent modulation of the number of carriers in the device channel [17].

Despite the already known advantages of ImmunoFETs, only potential variations that occur within the EDL (electric double layer) or Debye length (λ_D) from the surface of the FET can be detected by these devices because the counterions in the solution create an effect of electrostatic shielding. The Debye length is strongly dependent on the ionic strength of the solution. The majority of developed ImmunoFETs can only detect Ab-Ag interactions directly in solutions of low ionic strength (See in 2.1.5 a discussion about λ_D and strategies that are adopted to create detectable signals in ImmunoFETs).

Silicon nitride ISFET was used as ImmunoFET to detect tumor necrosis factor-alpha (TNF-α) in saliva [18]. TNF-α is a pro-inflammatory cytokine and represents a biomarker of heart failure. This ISFET was modified with anti-TNF-α antibodies, and using electrochemical impedance spectroscopy as a measurement strategy presented a limit of detection of 1 pg mL^{-1} [18]. SEGFET was used as ImmunoFET to detect

Dengue biomarker antigens (non-structural protein 1, NS1) [19]. For this purpose, gold electrodes modified with anti-NS1 antibodies were connected to a commercial MOSFET. The detection of NS1 antigens allows diagnosing Dengue early, even before symptoms, from the second or third day of infection. SEGFET detected these antigens in a concentration range of 0.25–5.0 $\mu g\ mL^{-1}$ in low ionic strength buffer solutions, suggesting indicating that the system is promising for the early diagnosis of Dengue [19].

Magliulo et al. fabricated OFETs in the configuration of solution-gated FET for the label-free detection of C-reactive protein (CRP). CRP is a serum protein and is a good serum biomarker of cardiovascular, inflammatory or cancerous diseases. To realize the ImmunoFET, poly-3-hexyl thiophene (P3HT) was deposited on an interdigitated electrode and modified with anti-CRP antibodies. The device operated from 4 pM to 2 μM CRP concentration range and achieved a limit of detection of 2 pM [20]. The same group used a similar OFET to detect procalcitonin, a biomarker used as an indicator of sepsis and to aid in administering antibiotics [21]. A label-free ImmunoFET was utilized for the detection of IgG antibodies. In this case, OFET is separated from the sensitive part (i.e., a gold electrode) in the EGFET configuration. A linear range from 0 to 10 $\mu g\ mL^{-1}$ was obtained with a limit of detection of 0.62 $\mu g\ mL^{-1}$ [22]. Wang et al. highlight other applications of OFETs as biosensors [23].

SiNWs have been widely used to fabricate ImmunoFETs to detect several disease-relevant biomarkers such as troponins [24, 25], proteins indicative of myocardial infarction, and some biomarkers indicative of cancer [26]. In the same way, the neuropeptide orexin-A, a biomarker of fatigue and cognitive performance, was detected using ZnO nanowire FETs [27]. SiNWs-TFET biosensors can provide considerable improvement in sensitivity and differentiate noise from specific protein binding signals due to the ambipolar effects of TFETs [28]. Gao et al. presented an immunosensor with marker-free detection of CYFRA21-1 up to 0.5 $fg\ ml^{-1}$ (~12.5 aM) using a highly responsive SiNW-TFET device, with a minimal subthreshold slope of 37 $mV\ dec^{-1}$ [28]. CYFRA21-1 is one of the most sensitive biomarkers for non-small cell lung cancer prognosis. In the study, a CYFRA21-1-sensitive SiNW-TFET biosensor was created by immobilizing a specific antibody on the SiNW surface [28].

We have recently shown that FETs formed by multilayers of rGO can act as ImmunoFETs. In this case, rGO was modified with papain as a biological recognition element and used to detect Cystatin-C antigen, an early chronic kidney disease biomarker [29]. gFETs are moving toward the development of new diagnostic systems, certainly driven by the COVID-19 pandemic. Zhang et al. showed that gFETs could act as ImmunoFETs to detect Severe Acute Respiratory Syndrome Coronavirus 2 (SARS-CoV-2) antigens. They proved the electrical probing SARS-CoV-2 spike protein using gFET containing immobilized anti-spike antibodies or ACE2 (angiotensin-converting enzyme 2) protein as biological recognition elements. The system detected the protein concentrations within 2 min with a limit of detection of 0.2 pM in 0.01 M PBS buffer solution. Still, for COVID-19 diagnosis, Seo et al. demonstrated that gFETs with anti-spike antibodies could detect the spike

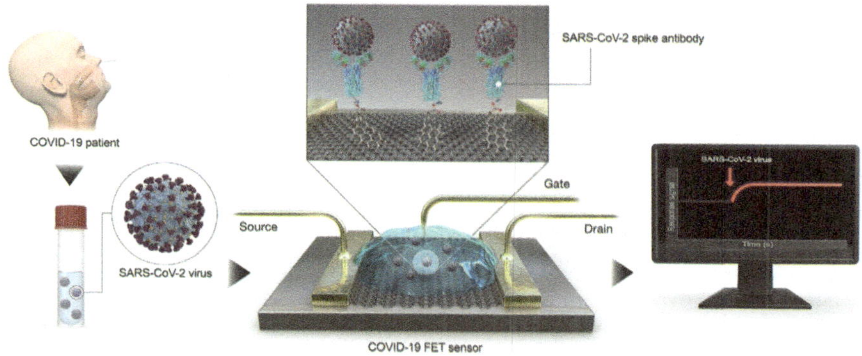

Fig. 2 Schematic diagram of COVID-19 FET sensor operation procedure. Graphene as a sensing material is selected, and SARS-CoV-2 spike antibody is conjugated onto the graphene sheet via 1-pyrenebutyric acid *N*-hydroxysuccinimide ester, which is an interfacing molecule as a probe linker. Reprinted with permission from reference [30]. Copyright (2020) American Chemical Society

protein of SARS-CoV-2 in real samples of patients, that is, nasopharyngeal swab specimens [30]. More specifically, the device could detect the SARS-CoV-2 spike protein as low as 1 fg/mL in PBS and 100 fg/mL clinical transport medium. In addition, this immunoFET presented a limit of detection of 1.6 pfu/mL in the culture medium and 2.42×10^2 copies/mL in clinical samples. Figure 2 displays the schematic diagram of this graphene ImmunoFET [30]. Furthermore, graphene nanoribbons FETs, first utilized for the detection of heart disease biomarkers, are being adapted for diagnosing COVID-19 at the company Nano DiagnosiX (Fayetteville, AR, USA) [31]. Shao et al. recently used SWNTs NanoFETs modified with anti-spike protein and anti-nucleocapsid protein antibodies to detect SARS-CoV-2 antigens. The ImmunoFET achieved a limit of detection of 0.55 fg/mL and 0.016 fg/mL for spike protein and nucleocapsid protein antigens, respectively. Also, the proposed ImmunoFET exhibited good performance in discriminating positive and negative clinical samples [32].

Considering a lot of known disease biomarker antigens and the increasingly mature of FET technology, ImmunoFETs are expected to reach the commercialization stage. This would be very important, especially in pandemic periods or in regions with endemic diseases such as Dengue, Zika, Yellow Fever, etc. Since FETs can be produced on a large scale and at an affordable cost, more diagnostic tests would be accessible to the population, allowing for a greater number of diagnoses at a lower price, especially in developing countries.

2.1.3 DNA-FETs

The detection of nucleic acids (NA) using the so-called DNA-FET has attracted significant interest because of the possibility of eliminating thermal blocks for

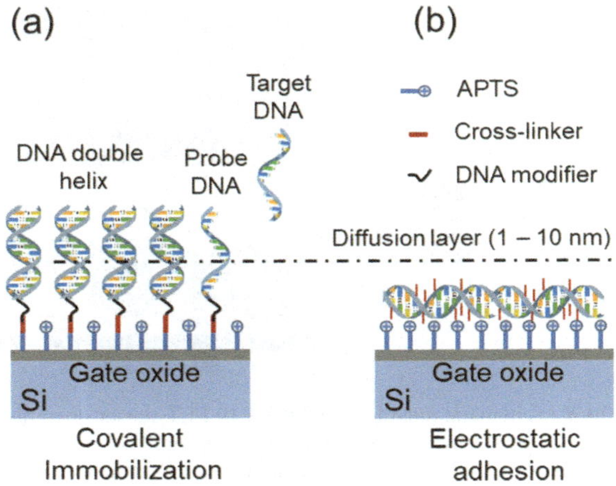

Fig. 3 **a** Probe DNA is covalently attached utilizing a cross-linker, in combination with target DNA, forms a DNA double helix. **b** DNA molecules are attached to the 3-Aminopropyltriethoxysilane (APTS) layer by electrostatic immobilization. Adapted from reference [34]

amplification and optics apparatus for detecting NA sequences and fluorescently labeled sequence-specific DNA probes or fluorescent dyes for labeling NA present in traditional technics [33].

A DNA-FET is fabricated by immobilizing a specific single-stranded capture DNA probes on the gate insulator. The complementary target probe hybridizes with high specificity on the immobilized probe. Because DNA is an intrinsically negatively charged molecule due to the presence of a phosphate backbone, the charge density creates a potential shift near the gate upon hybridization. Figure 3 illustrates two methods for immobilizing DNA "capture" probes [34]. The method shown in Fig. 3a is based on a DNA capture probe covalently immobilized on the surface. In this approach, a smaller portion of the target is within the EDL, but the hybridization is not severely hindered [34]. The process in Fig. 3b is an electrostatic adhesion. The target may be within the EDL in this method. However, the hybridization efficiency in this method is not very high because the target needs to wound around the probe [34].

The detection of DNA sequences is an essential tool in medicine and biology to detect specific mutations aiming at personal therapies and biological tracking, particularly in pandemic scenarios, like COVID-19. ISFET integrated with CMOS technology fabrication is gaining importance to DNA sequencing due to its inherent scalability, possible real-time detection, low-cost, and portable and integrated sensing devices development. In this way, Rothberg et al. reported a chip capable of sequencing a complete genome by synthesis based on hydrogen ions detection using an array of ISFET sensors [35]. The nucleic acid polymerization reaction releases hydrogen ions during nucleotide incorporation, which is responsible for changing the pH of each chip well (a pH variation of 0.02 for each single base incorporation). The sensor allows for faster readout speed, good cost–benefit, large-scale production, and lower cost per base-pair (*bp*) (~\$10 to \$100) when compared to second-generation platforms [35]. Nevertheless, the ISFET sensor cannot detect repetitive sequences of a single nucleotide that are incorporated in a single reading [35].

The technology Ion Torrent platform is considered a transition between second and third-generation platforms that produce 15 Gb of bases per run with 400 bp readings and run time of 2 h. The challenger for third-generation DNA sequencing, currently under active development, is the real-time sequencing of a single molecule, which requires integrated amplification reactions and molecule selection mechanism using nanopores [36]. Thus, Toumazou et al. described an integrated circuit based on CMOS design that integrates ISFET array sensors in an entire system-on-chip platform with simultaneous DNA detection and amplification, optimizing the classical PCR and loop-mediated isothermal amplification methods for pH detection in low buffering conditions [37].

NanoFETs are also utilized as DNA-FET. In these nanosensors, most often, the capture DNA probes are immobilized on the nanosized channel. By using CVD-grown monolayer MoS_2 films, Liu et al. fabricated highly sensitive DNA-FET to detect the peripheral blood of pregnant women over-expression of chromosome [38]. Gold nanoparticles (AuNPs) were deposited on the MoS_2 channel, and then the DNA capture probes were immobilized on the AuNPs. The sensor achieved an ultra-high sensitivity for the reliable detection of target DNA with a limit of detection of 0.1 fM. High specificity for 3-nucleotide polymorphism discrimination proved their potential for detecting chromosome 21 over-expression and the device achieving Down syndrome screening [38]. An ultra-sensitive ZnO-doped MoS_2 NWs FET was reported by Shariati et al. for hepatitis B virus (HBV) [39]. This HBV DNA-FET showed high selectivity by discriminating complementary DNA from non-complementary, and the mismatch one, two, or three bases oligonucleotides in blood serum and could detect the small molecules with a detention limit of 1 fM in a 0.5 pM to 50 mM linear concentration range [39].

Graphene has been attractive for constructed DNA-FET (or DNA-NanoFET) due to its unique properties. Han et al. reported a graphene DNA-FET for the real-time monitoring of loop-mediated isothermal amplification (LAMP) of DNA. They could detect viral DNA within 16.5 min, with a limit of detection of 2×10^2 copies/μL (or 10 fg/μL) [40]. A significant advance was achieved in the graphene DNA-FET to detect cancer-related microRNAs in undiluted human serum using a deformed monolayer graphene channel shown in Fig. 4 [41]. This device showed an ultra-high sensitivity in buffer (600 zM) and human serum samples (20 aM). These concentrations correspond to ~18 and ~600 NA molecules, respectively. The experiments and simulations showed that the nanoscale bending and deformations employed on FET fabrication increase the Debye length in ionic solution and decrease the screening effect, significantly enhancing sensitivity compared to flat graphene DNA-FET [41].

Recently, Li et al. reported a DNA-FET for the rapid and unamplified identification of COVID-19 in human throat swab specimens [42]. The DNA-FET was fabricated using AuNPs-modified rGO with immobilized complementary phosphorodiamidate morpholino oligos (PMO) probe. PMO is an uncharged DNA analog [42]. Using this DNA-FET, the analyses of RNA extracts from 30 real clinical samples were in high agreement with the RT-PCR method [42]. Specific response to COVID-19

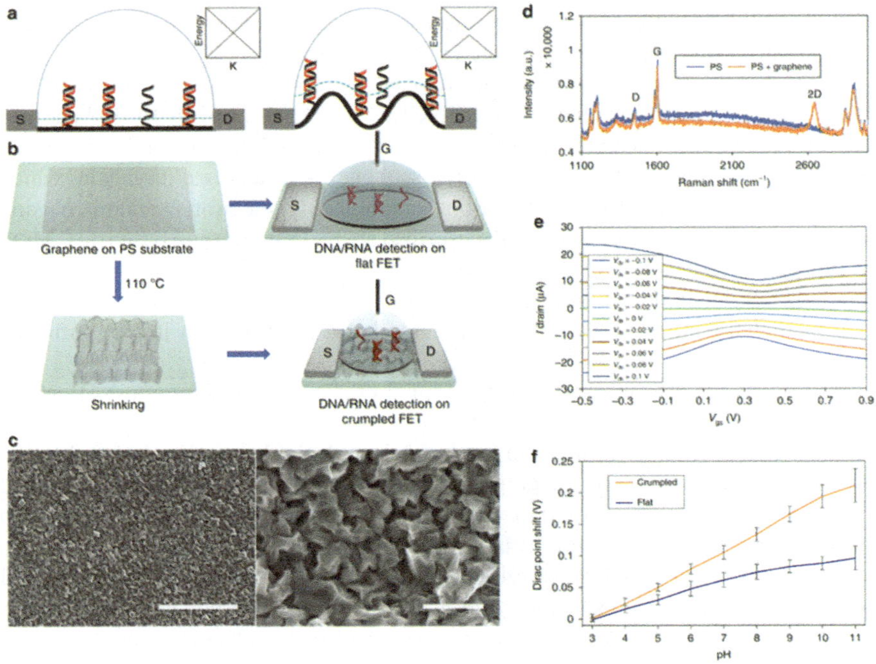

Fig. 4 Scheme and characterization of flat and crumpled graphene FET biosensor. **a** Cross-sectional scheme of the flat (left) and crumpled (right) graphene FET DNA sensor. Probe (black) and target (red) DNA strands are immobilized on the surface of graphene. The blue dot lines represent Debye length in the ionic solution, and the length is increased at the convex region of the crumpled graphene, thus more area DNA is inside the Debye length, which makes the crumpled graphene more electrically susceptible to the negative charge of DNA. The inset boxes represent qualitative energy diagram in K-space. Graphene does not have intrinsic bandgap. However, crumpled graphene may open bandgap, which is discussed in the later section and supplementary table. **b** Fabrication of FETs and experimental process flow. Graphene on pre-strained PS substrate was annealed at 110 °C to shrink the substrate and crumple the graphene. Then urce and drain electrodes were applied, and solution-top gate was used. In case of flat graphene FET, the annealing process was omitted. **c** SEM images of crumpled graphene. The scale bar is 5 μm (left) and 500 nm (right). **d** Raman spectroscopy of crumpled graphene and PS substrate. **e** Charge transfer characteristics of the fabricated crumpled graphene FET. Vgs vs Ids (bottom) with the variation of Vds graphs showed shift in the Dirac point. **f** Dirac point shifts of the FET sensor plotted as a function of pH values. n = 5, mean ± std. Reprinted with permission from reference [41]. Copyright (2020) Springer Nature

patient samples can be achieved within 2 min by real-time measurement using this sensor, a sample preparation time of 30 min, totaling 32 min of complete molecular analysis. Although there is a simple laboratory preparation for RNA extraction, the study brings a significant advance in molecular label-free detection and free from amplification of genetic material. Among the advances brought by Li et al. study in detecting viral RNA are PCR-free direct detection without pre-amplification, the low limit of detection (0.37 fM), and the differentiation between SARS-CoV-2 and SARS-CoV [42].

The search for the detection of molecular label-free in samples of liquid biopsies, including extracellular vesicles (EV) and microRNA extraction, can be achieved with the microfluidic systems conjugated with the FETs. In this line, a sapphire-based FET biosensor was reported for label-free detection of DNA hybridization using graphene and a microfluidic system [43]. The graphene FET array was directly fabricated on the substrate using CVD, i.e., without metal catalysts synthesis. Using a homemade microfluidic system, the device could detect a concentration as low as 100 fM DNA detection sensitivity [43]. A DNA-FET integrated with a microfluidic system was recently developed to quantify breast cancer biomarkers concentrations (microRNA-195 and microRNA-126) [44]. This study showed that DNA-FET biosensors could detect concentrations of the microRNA-195 and microRNA-126 down to 84 and 75 aM, respectively. The microfluidic chip was designed to extract EVs from plasma using anti-CD63 beads at the EV extraction module and EV-encapsulated microRNAs (EEMs) using magnetic beads overlay with amine-modified cDNA probes (at the microRNA extraction module) before microRNA detection on a DNA-FET sensing module. Pneumatically driven microfluidic devices automated the entire process and could quantify microRNA-195 and microRNA-126 within five hours, showing its potential for early-stage breast cancer diagnosis [44].

TFET-based biosensors hold promise for detecting label-free DNA. Recently, Priyanka et al. showed, through simulations, the feasibility of the field-effect transistor based on inversion and tunneling for DNA detection [45]. In this study, an n-channel TFET has been proposed for the detection of label-free DNA. The nanogap is shaped like a suspension bridge that provides positioning and moving the biomolecules for dielectric modulation. The structure allows the biomolecules to be present on both the source and the drain side of the nanogap. The n-channel TFET showed better sensing capability compared to n-channel IMFET (inversion mode field-effect transistor). The simulation results are promising, and the n-channel TFETs arouse great interest for future research regarding the change in the permittivity, orientation angle, and negative charge density of DNA biomolecules in high-sensitive DNA biosensors [45].

Despite the great advantages of DNA-FETs, questions like: instability in aqueous conditions, less sensitivity of some systems, the use of conventional Ag/AgCl electrodes for gating, simultaneous amplification, and label-free detection of DNA hybridization still have been the focus of research. As mentioned, significant interest and efforts are aimed at identifying common and rare single-nucleotide variations. When these challenges are fully overcome, it is expected that detecting specific NA sequences and their variants rapidly using DNA-FETs will improve disease diagnostics.

2.1.4 New Trends: CRISPR FETs

Regularly clustered interspaced short palindromic repeats (CRISPR) and CRISPR-associated proteins (Cas) have recently emerged as a new technology for NA detection. Briefly, the CRISPR-Cas system is a prokaryotic adaptive immunity mechanism used to cleave attacking NA. For a complete overview of the molecular biology of this system, see the references [46, 47]

Due to its high selectivity, the CRISPR-Cas system is already being used as a biological recognition element in biosensors [48]. There are only two studies in the literature so far regarding FET biosensors using CRISPR-Cas technology, both using graphene. In the first study, the CRISPR-Cas9 system was immobilized on gFET for the label-free detection of unamplified target genes [49]. gFETs were obtained from a commercial foundry with Cardea Bio company [50]. Combining the gene-targeting capacity of CRISPR-Cas9 with the sensitivity of gFET enabled the detection of a target sequence contained within intact genomic DNA within 15 min. The clinical utility was proved the detect Duchenne muscular dystrophy-associated mutations and reached a limit of detection of 1.7 fM without amplification [49]. In the second study, the group demonstrated the detection of single-nucleotide polymorphisms (SNPs) in DNA without target amplification [51], using the same gFET from Cardea Bio, named in that study of "CRISPR-SNP-Chip" [51] (Fig. 5). The authors detected SNPs in target genes in two human diseases: sickle cell disease and amyotrophic lateral

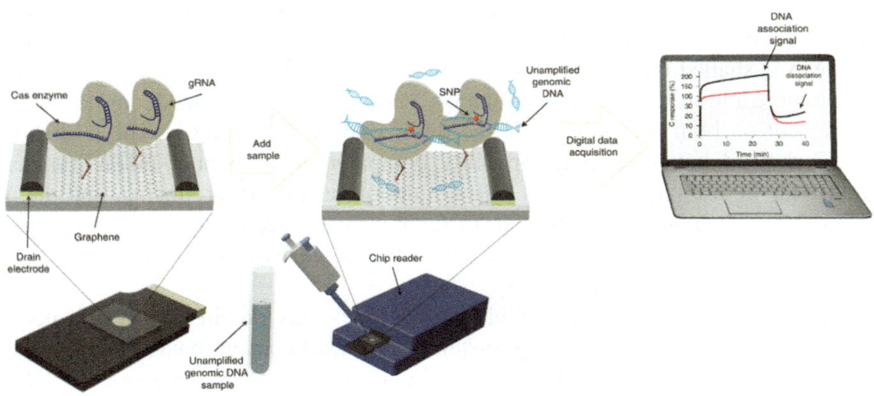

Fig. 5 CRISPR-powered gFET for amplification-free detection of single-nucleotide mutations. SNP-Chip, the next generation of CRISPR-powered gFET, with expanded monitoring of multiple electronic parameters, can detect single-nucleotide differences within unamplified DNA samples. Through the CRISPR–Cas9 ribonucleic protein complex on its surface, this technology can digitally detect SNPs without labels or amplification. The target-specific gRNA is designed to target single-nucleotide mutations relevant to two human disease models: SCD and ALS. Through real-time, multiparameter, and digital data acquisition, SNP-Chip can discriminate between unamplified genomic DNA samples in 40 min. Reprinted with permission from reference [51]. Copyright (2020) Springer Nature

sclerosis, suggesting a great contribution to medical diagnosis and basic research [51].

It is expected that the CRISPR-Cas technology combined with FETs transducers allows the next generation of biosensors to detect nucleic acids, producing point-of-care-type detection systems without a need to use trained personnel and expensive equipment.

2.1.5 Debye Length Screening in FETs

The applied gate voltage through the electrolyte/channel interface will induce opposite charges forming a double-charged layer in the electrolyte solution. Then, the FET response depends on the charge distribution near the sensing surface (FET channel). In solution, there is a unique relationship between the electrostatic potential of the ionic array and the charges distribution in that array (in a specific way) [52]. Therefore, we can approximate the potential change at the sensing surface by a parallel combination of two capacitances: a double-layer capacitance (C_{DL}) and a sensor FET capacitance (C_{FET})—the C_{FET} capacitance comprises the oxide and depletion capacitances [53]:

$$\psi_0 = \frac{\Delta\sigma_0}{C_{\mathrm{DL}} + C_{\mathrm{FET}}} \tag{1}$$

In this case, the pH sensitivity of oxide interfaces (oxide capacitance, C_{ox}) is described using the site binding model incorporating by the C_{DL} capacitance [53].

The Poisson equation gives this charges distribution, which can be derived from the Gauss' law of electromagnetism. In this case, assuming a conservative field, one can relates the charges distribution with potential gradient. Then, starting from the Poisson equation, we can get a better description of the double-layer formation (considering one dimension) [54]:

$$\nabla^2\phi(x) = -\frac{\rho(x)}{\varepsilon} \tag{2}$$

where $\rho(x)$ is the density of charges and ε is the dielectric constant of the medium.

Assuming a fixed charge density, near the channel surface, and a fluctuation in $\rho(x)$ due to the thermal movement of the ions in solution (obeying the Boltzmann distribution using the statistical thermodynamics), we can approximate the potential by a linear approximation in terms of the first-order Taylor expansion. Debye proposed this approximation to small potentials comparing with thermal energy, obtaining [54]:

$$\nabla^2\phi(x) = \frac{\sum_j C_j^0 q_j^2}{\varepsilon K_{\mathrm{B}} T}\phi(x) - \frac{\rho(x)'}{\varepsilon} \tag{3}$$

where C_j^0 is the mean ion concentration, q_j is the charge, ε is the gate dielectric, K_B is Boltzmann's constant, T is the absolute temperature (in K), and $\rho(x)'$ is the fixed density of charge.

Disregarding the charge carriers dimensions, the detectable distance from the surface of the channel, i.e., the average penetration distance of the electric field, is given by the so-called Debye length (λ_D) formation [54]:

$$\lambda_D = \sqrt{\frac{\varepsilon K_B T}{2 N_A e^2 I}} \tag{4}$$

where ε is the gate dielectric, K_B is Boltzmann's constant, e is the elementary charge, T is the absolute temperature, N_A is the Avogadro's number, and I is the ionic strength of the solution.

Observe that the λ_D increases as the solution temperature (T) increases and decreases as ionic strength (I) increases. For example, the free carriers in metal are huge, and therefore, λ_D is very small, so the electric field hardly penetrates it. Thus, for an affinity biosensor like an antigen binds to an antibody immobilized on the sensitive layer of the FET, a change in the double-layer capacitance is expected if the binding distance is within the Debye length (see Fig. 6). In addition, the greater the ionic strength of the electrolyte, the greater the shielding effect caused by the distribution of the ions. Then, the ionic concentration of the sample has to be near an ideal electrolyte (~mM). Stern et al. discuss the importance of a solution for optimal sensing in nanowire FETs for complementary single-stranded DNA detection [55]. So, the Debye length determines the biological probe and, under physiological conditions, the effective sensing distance is given by $\lambda_D \approx 1$ nm (e.g., IgG 15 nm) [56].

Taking PBS buffer (phosphate buffer saline) as the body fluid mimetic system, the buffer has an ionic strength of 150 mM, representing a Debye length of ~0.7 nm. Considering now that an immobilized antibody molecule is 10–15 nm, it is clear that antibody–antigen interactions will only be detectable by ImmunoFETs in solutions of low ionic strength [58]. Thereby, additional steps such as dilution or desalination when dealing with a real sample (serum or blood plasma, for example) are required. In addition to increasing the assay time, diluting the sample implies developing ImmunoFETs with a low limit of detection. On the other hand, desalination can lead to lower charges on the biomarker molecules (antigen or antibody), generating a lower output signal in the devices [59].

To overcome the Debye length screening, some strategies have been adopted besides reducing receptor size or decreasing the ionic strength of the solution. For example, the use of label molecules in ELISA-based ImmunoFET was reported by Jang et al. [60]. The authors used the catalyst effect of alkaline phosphatase to induce Ag precipitation and overpass the Debye screening. Other approaches are: improving materials' morphology like the use of NanoFETs with concave morphology [53]

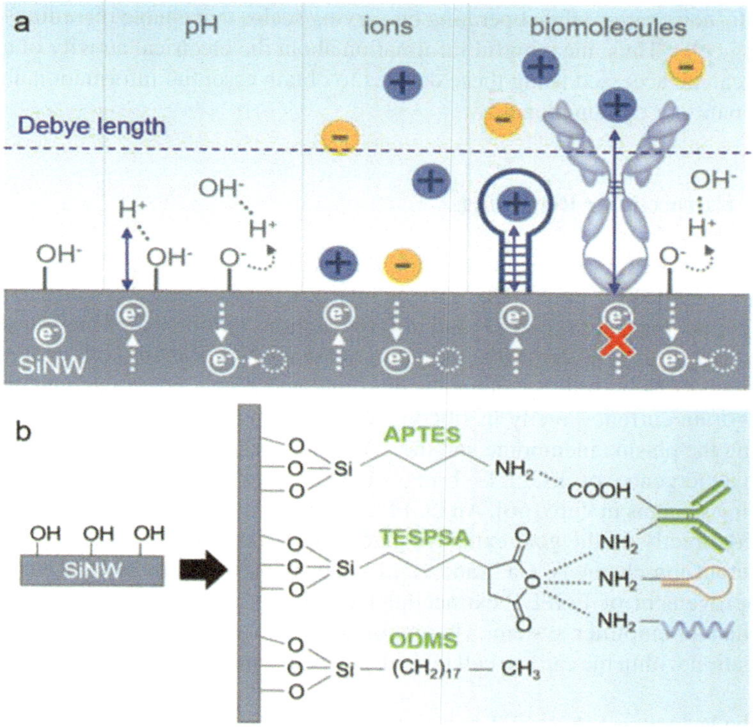

Fig. 6 **a** Importance of specificity and Debye layer. Only charges under the Debye length limit will have influence on FET conductivity. When there are no receptors (left area), pH and the presence of charged molecules (either specific or unspecific) will determine charge carrier movement. In the presence of bioreceptors, only specific targets will bind, avoiding interference on unspecific charges. However, using large receptors, the "caught" charges will remain above the Debye length, showing no conductivity modulation. pH still interacts, though. **b** Examples of SiNW surface modification using silanes containing different functional groups where bioreceptors (antibodies or amino-modified aptamers or single-stranded DNA molecules) can be immobilized. Inert silanes can also be attached for chemical passivation. Reprinted with permission from reference [57]. Copyright (2019) Wiley Online Library

or permeable and porous polymers to the channel surface [61], and use of high-frequency signal (AC) to avoid the double-layer formation (typically a sinusoidal signal of 10 mV of amplitude and frequency between 1 kHz and 1 MHz) [62].

2.2 *Cell Interfaces*

FET transducers have achieved advances in electrophysiology as an advanced method for measuring the electrical potentials of cells and tissues. FETs maintain very good signal-to-noise ratios while operating on varying scales that enable recording signals in living cells. Thus, meaningful information about the electrical activity of a tissue or cell can be accessed using these devices to obtain essential information about its functionality or dysfunctionality.

2.2.1 Extracellular Recordings

The research in FETs for quantifying the electrical activity of cells started in the first studies on the ISFET to recorded extracellular ion pulses measured near the muscle of the guinea-pig taenia coli [63] and in vivo reactions of the flexor tibialis muscle of a locust [64]. Later, in the 1990s, a neuron of the leech, a Retzius cell, was attached to the gate of FET [65]. Spontaneous or stimulated action potentials modulate the source–drain current directly in silicon [65]. This study described the junctions by coupling the plasma membrane and the gate insulator [65].

The emergence of EGFETs brings another way for long-term extracellular recording systems in vitro [66]. An EGFET system tested using rat cardiac myocytes cultured directly on the gold extended gate surface can distinguish the contribution of different ion channels, Ca^{2+} and K^{+}, by the shape of the recorded signals [66]. An improvement of EGFET extracellular recording employed a configuration of a 64-channel amplifier system. This multiplexed biosensor could detect the signal propagation within the cardiac cell layer using all channels at the same time at 10 kHz [67].

In the last years, SiNW-FET has gained ground in extracellular recordings detection. Shang et al. reported a SiNW-FET coated with polyvinyl chloride containing a valinomycin membrane to detect the efflux of K^{+} from chromaffin cells [68]. This biosensor presented could detect K^{+} in a range of 10^{-6} to 10^{-2} M with good sensitivity and selective [68].

Dopamine (DA) is a neurotransmitter that plays several crucial roles in the brain. Its dysregulation is associated with various diseases, such as Parkinson's disease. Nonetheless, the concentration of DA is extremely low in patients and is difficult to detect using existing conventional methods [69]. Li et al. developed a SiNWs-FET for ultra-sensitive and selective detection of DA release from Living PC12 cells under hypoxic stimulation [70]. This biosensor proposed consists of multiple-parallel-connected modifying DNA-aptamers (MPC aptamer/SiNW-FET) that show higher sensitivity than a traditional single-channel SiNW-FET. The results indicated the specificity of the biosensor in discriminating DA from other molecules [70].

Recently, Park et al. used the same strategy with specific aptamers to detect DA using carboxylated polypyrrole nanotubes (CPNTs) FETs [71]. The DA concentration by modulating the FET gate potential showed highly sensitive detection. In the DA release assay in PC12 cells, exogenous DA levels were successfully quantified, according to the scheme displayed in Fig. 7. The detection characteristics of the CPNTs-FET aptasensors were compared with high-performance liquid chromatography (HPLC) standards. According to Park et al., these CPNT-FETs provide fast and

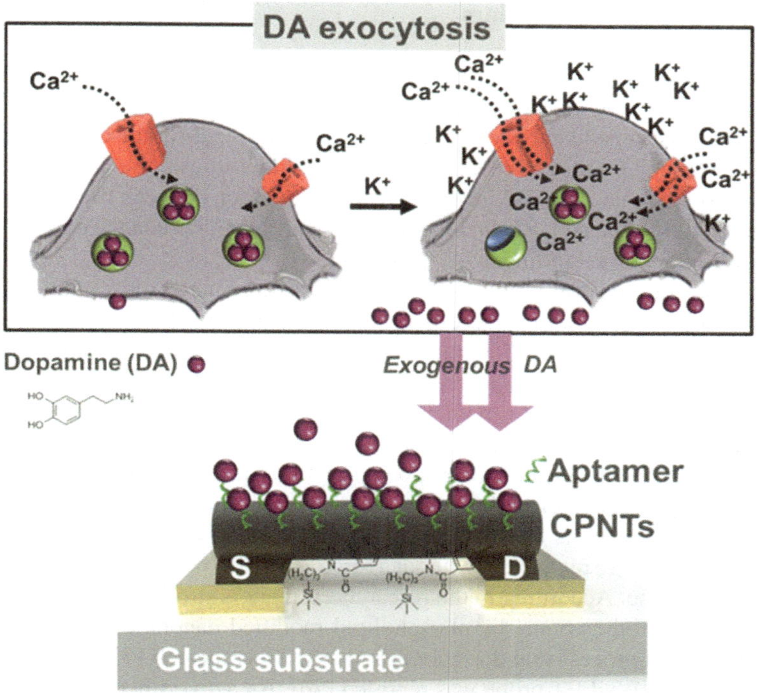

Fig. 7 Schematic illustration of cell exocytosis for DA release from PC12 cells via rapid Ca^{2+} reflux accelerated by K^+ ions (upper) and liquid-ion-gated FET aptasensors using aptamer-conjugated CPNTs for exogeneous DA detection (down). Reprinted with permission from reference [71]. Copyright (2020) Springer Nature

efficient detection and can be applied to identify genetic diseases linked to neurons, such as Parkinson's disease [71].

A midway step toward in vivo measurements is ex vivo recordings, for example, from heart tissue. In this way, Kireev et al. fabricated gFET on different substrates to monitor cardiomyocyte activity [72]. When manufactured on flexible polyimide substrates, the devices achieved large values of transconductance and mobility, up to 11 mS V^{-1} and 1750 cm^2 V^{-1} s^{-1}, respectively [72]. Recently, Kyndiah et al. reported an electrolyte-gated OFET sensor used to record the bioelectrical signal of spontaneously beating embryonic cardiomyocyte cells [73]. The performance of this OFET seems to be unchanged during the full experimental duration of ten days. Cardiac action potentials of 40 μV were recorded at a frequency of 0.65 Hz equivalent to beating/contraction of the cardiac cells [73]. The effect of pharmaceutical drugs such as Norepinephrine and Verapamil on the activity of the cardiac cells was successfully demonstrated using the OFETs [73].

2.2.2 Intracellular Recordings

The significant advances in electrophysiology have been the measure of the electrical potentials within a cell. The standard gold method consists of inserting an electrolyte-filled glass micropipette into the cell, providing a high signal-to-noise ratio [74]. Two factors are crucial for measurement accuracy: the dimensions of the micropipette, as small as possible, and the low impedance at the interface between the micropipette and the interior of the cell. Reconciling these two factors has been the current search for intracellular cell recordings [74].

In the last decade, the use of SiNWs-FETs has dominated most applications to record electrical potentials inside cells. As the performance of FETs is not impedance dependent, these devices can be much smaller than traditionally used micropipettes and microelectrodes. Duan et al. reported the ability of SiNWs-FETs to record the spontaneous intracellular transmembrane potential in chicken embryonic cardiomyocyte cells. The approach consists of a SiO_2 nanotube integrated into the gate of a FET that can penetrate the cell membrane, being less invasive and capable of recording the electrical potentials of both the transmembrane and the cell cytosol [75]. The relevance of this work is due to the sequential nature with which the intracellular potential signals were registered, without a force applied to penetrate the nanotubes in the cell membrane. According to the researchers, there is strong evidence that penetration of the cell membrane by phospholipid-modified nanotubes is a biomimetic and spontaneous process that does not negatively affect the cell [75].

A similar SiNWs-FET probe was reported [76]. An active silicon nanotube transistor (ANTT) was synthesized, which allows intracellular recording at high resolution and measurements of intracellular action potentials of individual cells in a multiplexed array. Figure 8 shows the design, synthesis steps, fabrication, and testing of this hollow needle-shaped active silicon nanoprobe configured as the gate of a FET [76]. ANTT probe interface studies in spontaneously beating cardiomyocytes have demonstrated recording intracellular action potentials with stable amplitudes, measurements similar to those reported by other electrophysiological techniques. Furthermore, the device showed that action potentials could be registered with nanotubes with internal diameters as small as 15 nm [76].

Zhao et al. demonstrated that ultra-small probe arrays in SiNWs-FETs is capable to record intracellular action potentials from primary neurons and other electrogenic cells [77]. The study strategy combines tip geometry and gate size, allowing the recording of intracellular action potentials up to 100 mV. Probes fabricated from p-type SiNWs can be used for intracellular recordings in cultured primary neurons and human cardiomyocytes. The reduction in channel length showed a significant correlation with increases in maximum measured action potential amplitudes. The results indicate that these SiNWs-FETs with biomimetic modification of phospholipids can directly access the cell's interior and produce an accurate record of the intracellular potential [77].

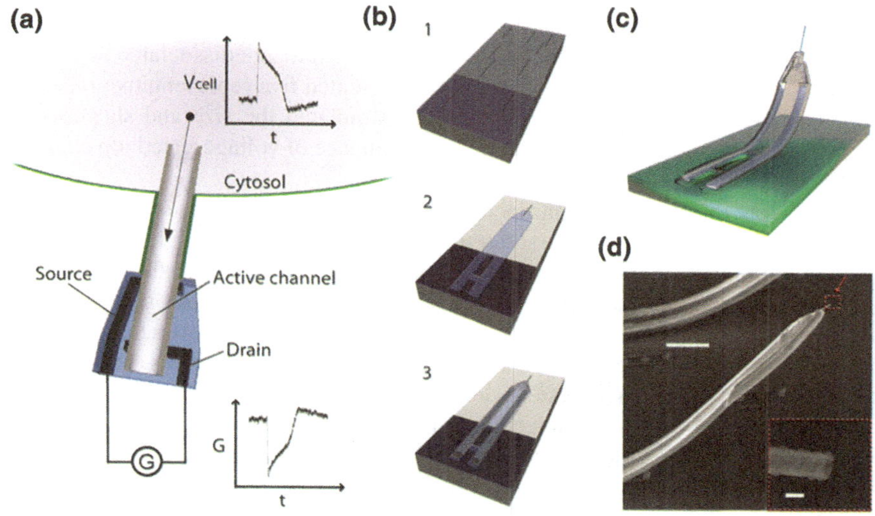

Fig. 8 Principle and fabrication of the ANTT probe. **a** Schematic view of an ANTT probe inserted into a cell and recording an intracellular action potential (Vcell vs. time, t) as a conductance (G) change in the active FET region between S/D contacts. Sensitivity to voltage changes from the external extracellular environment is effectively eliminated by SU-8 passivation of the nanotube region around the S/D contacts. The nanotube is shown as a halfcylinder for clarity. **b** Overview of the steps used for ANTT probe fabrication: (1) Transfer of germanium/silicon core/shell nanowires (Ge/SiNWs) to a SU-8 layer that was deposited and prebaked on a sacrificial layer (colored silver). (2) Registration of positions of Ge/Si NWs and definition of the bottom SU-8 layer. (3) Definition of S/D metal contacts followed by the top SU-8 passivation layer. Final etching of the sacrificial layer and Ge core yields the Si ANTT probe. **c** Schematic of the completed ANTT probe following release from the substrate. **d** SEM image of an ANTT probe. Scale bar, 10 μm. Inset, zoom of the probe tip from the dashed red box. Scale bar, 100 nm. Reprinted with permission from reference [76]. Copyright (2012) American Chemical Society

2.3 *Tissue Interfaces*

As seen in the previous sections, several FET devices were used to record intracellular and extracellular signals. In most cases, these devices are designed for recording signals from single cells or cell cultures. On the other hand, tissues are three-dimensional networks of cells with specific functions. For this reason, the position, shape, and size of the FET device respectively to the tissue under study are determining factors to the coupling [2]. In this section, we highlight some applications of FETs with biological tissue.

The pioneering studies by Bergveld had already shown the interface with biological tissues for measuring potentials near muscle tissues of guinea pig [63] and locust [64]. Jobling et al. used an array of nine gold electrodes in which each was connected to an individual FET for recording extracellular potentials variations on slices of tissue rat's brain [78]. Sprossler et al. also used a FET array for recording signals from rat cardiac muscle cells. The array was composed of 16 p-type FETs with non-metallized gates. Heart tissues of rats were minced, dissociated and put on the devices in a serum medium. The authors presented five representative measurements of the various types of couplings and simulated the size and shape of the recorded extracellular signals, indicating the influence of voltage-gated ion channels on signals [79].

OFETs are interesting for interfacing with cells and tissues because, among other features, polymeric materials are softer than inorganic ones and have the possibility of biocompatibility. Generally, the strain mismatch between the tissue and the OFET is very low. This reduces the inflammatory response of the surrounding tissue for in vivo applications [80]. Cramer et al. fabricated a pentacene OFET for extracellular stimulation and recording of activity of stem cell-derived neuronal networks. They showed the adhesion of murine neural stem cells on top of pentacene films without an additional layer of cell-adhesive molecules. OFET signal (i.e., current versus time) was monitored during the differentiation of stem cells and the formation of the neuronal network [81]. Au nanoparticles onto pentacene film were applied as OFET channel by Desbief et al. for interfacing with neurons [82]. As reported, human neuroblastoma stem cells were adhered, grown, and differentiated into neurons on de devices. The device simulates short-term neuronal plasticity (STP), a phenomenon observed in neuronal aggregates, due to the charging/discharging of gold nanoparticles [82].

Following the evolution of FET devices, it is known that sensors and biosensors based on NanoFETs have effects of size confinement give an exceptional sensitivity. Interfacing these devices with tissues is not different. Qing et al. manufactured SiNWs FET arrays on a quartz substrate to map neural circuits in acute brain slices. In that study, the transparent substrate enabled imaging of individual cell bodies and identifying areas of healthy neurons [83]. SiNWs FET arrays fabricated on planar and flexible polymeric substrates were also interfaced with cardiac tissue. It was shown that the conductance variations are synchronized with the beating heart signals [84]. Cheng et al. fabricated suspended gFETs for recording the beating heart tissue of rats. Figure 9 displays the 3D device scheme, the circuitry model of the graphene–tissue interface, and the recorded signals. Note that the suspended graphene yields an increase in signal amplitude and a decrease in noise [85].

It is expected that the true interface of FETs devices with biological tissues will be achieved, that is, within the body as non-invasively as possible and with adequate biocompatibility. In this way, gFETs seem adequate for in vivo interfacing applications because FETs can be fabricated in biocompatible flexible substrates. We can highlight some studies that move in this direction: Blaschke et al. perform experiments for in vivo mapping in the brain of anesthetized rats using flexible gFETs [8]. Masvidal-Codina et al. also use gFET arrays on flexible substrates to map cortical spreading depression in rats [86]. Furthermore, Hébert et al. shown that graphene ETs are efficient transducers for micro-electrocorticography [87].

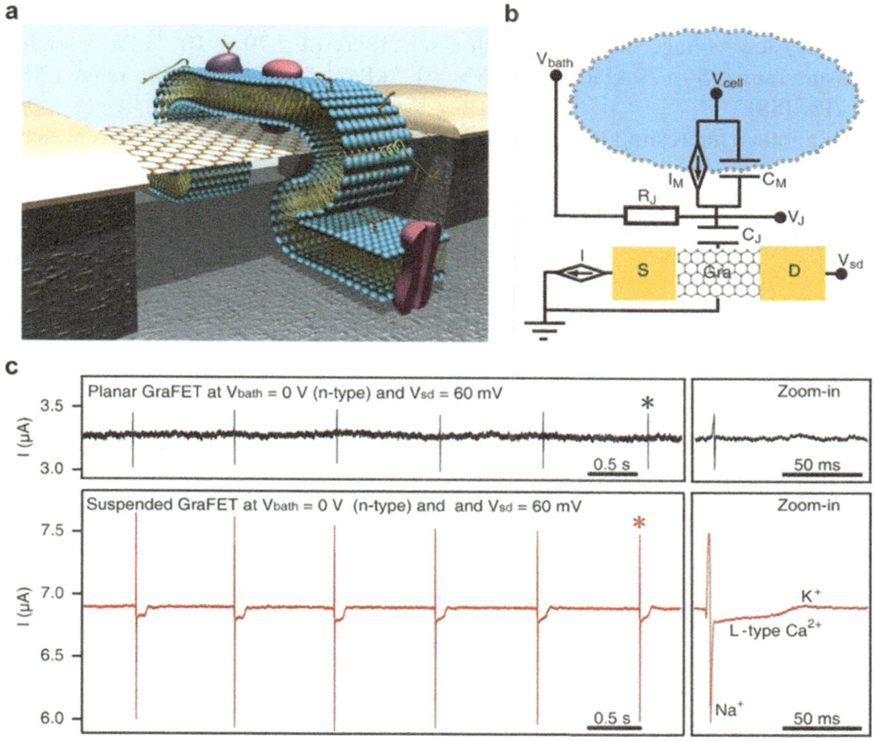

Fig. 9 Graphene transducer for extracellular recordings. **a** The schematic of a 3D graphene–biology interface. **b** The circuitry model of graphene–biology interface. **c** The recorded current signals from a heart by a graphene transducer, with an active area of 17.5 μm^2, before (black) and after suspension (red), respectively. The right panels are the zoom-in views of single eLFPs (extracellular local field potentials) at the time indicated by the stars on the current traces of the corresponding left panels. Reprinted with permission from reference [85]. Copyright (2013) American Chemical Society

2.4 *Others Biomedical Applications*

As shown in previous sessions, FET-based devices have emerged as tremendous potential in the biomedical area, and most researches are directed toward the development of biosensors. This section shows other biomedical applications of FET devices beyond the biosensors field, which capture some relevant insights for modern applications.

Wearable and biocompatible sensors with new features can emerge as a promising alternative to traditional stiff silicon semiconductors. In this manner, biomimetic hierarchical structures have been proposed using OFET sensors due to their inherently stretchable and flexible properties, simplicity, low-cost large-area fabrication, and the possibility to be tuned through chemical design. For example, Yogeswaran et al. reported a dynamic pressure sensor based on graphene in a SEGFET configuration aiming applications in electronic skin (eSkin) or prostheses [88]. The sensor operating at a very low voltage of 5.0 mV with a sensitivity of 2.70×10^{-4} kPa^{-1} for low pressure range (5 to 20 kPa) and 7.56×10^{-4} kPa^{-1} for high pressure range (20 to 35 kPa) [88].

In another interesting application, Kim et al. [89] emulated the afferent nerve system using a cluster of pressure sensors combined with an organic ring oscillator and an ion gel-gated transistor (synaptic transistor). The artificial afferent nerve based on flexible organic electronics was connected to the biological efferent nerve of a Blaberus discoidalis (discoid cockroach) to stimulate artificial leg contractions. Thus, the proposed systems have applications in neurorobotics and neuroprosthetics systems, particularly in brain–machine interfaces [89]. Feili et al. fabricated a flexible OFET sensor in a polymer substrate (polyimide) using pentacene as an active layer and sputtered gold as the electrodes (gate, drain, and source) [90]. The sensor was encapsulated using biocompatible parylene C polymer aiming at biomedical micro-implants. In particular, the sensor was proposed to be used as neural prostheses to restore body functions after paraplegia [90]. Figure 10 shows an example of an OFET mimicking a biological synapse by the associative response of a certain input signal

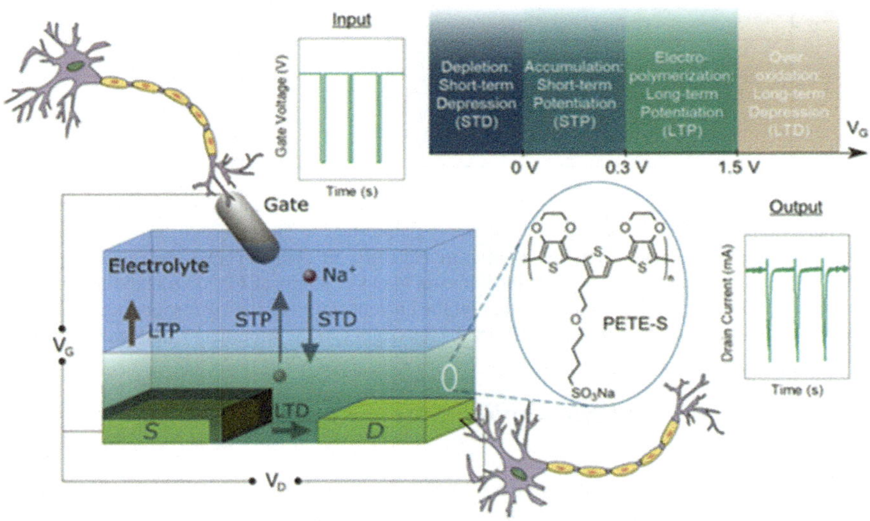

Fig. 10 Synaptic transistor. Sketch of the organic electrochemical transistor, formed by electropolymerization of ETE-S in the transistor channel. The electrolyte solution is confined by a PDMS well (not shown). In this work, we define the input at the gate as the presynaptic signal and the response at the drain as the postsynaptic terminal. During operation, the drain voltage is kept constant while the gate is pulsed. Synaptic weight is defined as the amplitude of the current response to a standard gate voltage characterization pulse of −0.1 V. Different memory functionalities are accessible by applying gate voltage spikes at different ranges or timing intervals. ETE-S is sodium 4-(2-(2,5-bis(2,3-dihydrothieno[3,4-b][1,4]dioxin-5-yl) thiophen-3-yl)ethoxy)butane-1-sulfonate monomer, PETE-S is the electropolymerization of ETE-S and PDMS is poly(dimethylsiloxane). Reprinted with permission from reference [91]. Copyright (2019) John Wiley and Sons

capable to short- and long-term memory effects and working as a synaptic transistor [91].

Other exciting applications of FET devices in the biomedical field are imaging (for cell activity monitoring, drug testing, and tissue engineering), FET-based dosimeters, and DNA sequencing based on ISFETs and multisensory systems used in electronic noses and tongues. We highlight some examples: Zoumpoulidis et al. fabricated a stretchable and biocompatible 2D array (matrix of 4 by 4) of CMOS-compatible ISFET sensors used in the biomedical imager. The proper functioning of the sensors requires electrical flexible interconnects, and the biocompatibility was possible by a protective encapsulation using parylene C polymer [92]. Zeng et al. fabricated a platform with high spatial resolution ion-sensing based on an array of 128 $\times$ 128 CMOS ISFETs for real-time ion imaging. The proposed system presents a frame rate of 3000 fps capable to detect the diffusion of ions in water at 60 fps [93]. Zeidell et al. developed a flexible OFET as a radiation dosimeter (energies between 0 and 6 MeV and radiation doses between 0.1 and 10 Gy) to be placed directly on a patient's skin in cancer diagnosis and treatment [94]. The deformable nature of the sensor permits its accommodation to the curvature of the skin surface. The authors also proposed a model to explain the sensor operation based on the interplay between charge photogeneration and trapping upon X-ray exposure [94]. Tang et al. reported using NanoFET based on SWCNTs for radiation dosimetry applications (ionizing energy of 6 MeV therapeutic x-rays and exposure to over 1 Gy) [95]. The study demonstrates the possibility of the miniaturization ($\sim$ nm) and integration of the X-rays dosimeters [95].

Electronic tongues and noses (respectively, e-tongues and e-noses) are sensors based on global analysis that can mimic the human sense of taste and smell. FETs transducers are also applied in this field. Figure 11 shows a sensor array of 20 OFETs based on metalloporphyrin receptors in a fully integrated e-nose system [96]. The chemical synthesis of organic compounds provides a wide variety of materials to be applied in OFETs. In addition, the organic nature of these materials has more compatibility with biomolecules allowing applications in biosensing platforms and biocompatible devices. Among the OFETs, biocompatible FET platforms for implantable devices, flat transistors, and bio-integrated wearable systems are the most promising candidates for fully integrated systems with the patient body in continuous monitoring.

3 Final Remarks and Challenges

Since Bergveld, considerable progress has been attained in FET-based devices to be applied in the biomedical area. In particular, the COVID-19 pandemic caused pressure to develop fast, disposable, low-cost, portable, and reliable bioanalytical sensors for population-wide screening. In this regard, developing FET devices enable extensive integration by microfabrication methods. The microfabrication processes can also be applied to micro-implants aiming at biomedical electrodes to probe cellular

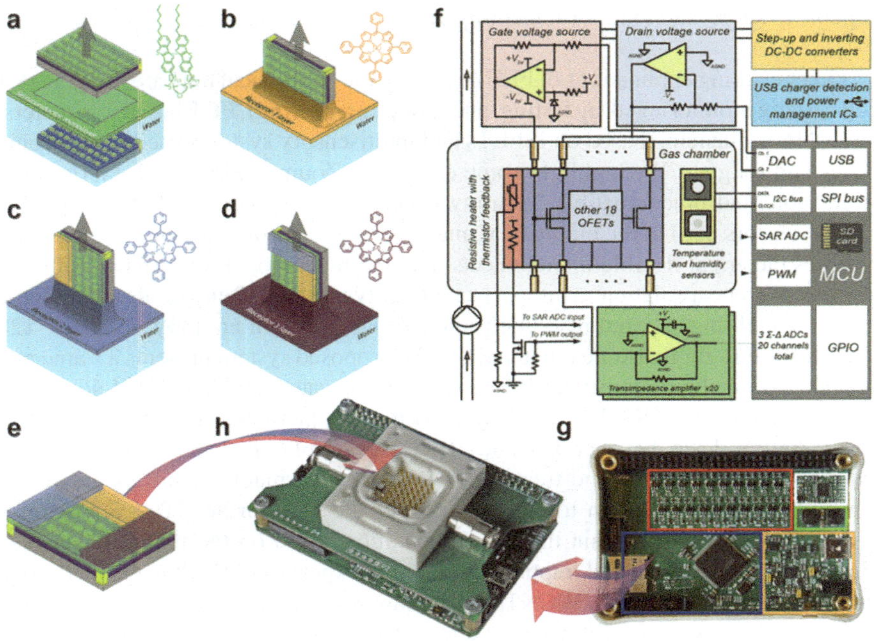

Fig. 11 Schematic representation of the sensor array fabrication process and the electronic nose images. **a** Langmuir film deposition of organic semiconductor D2-Und-BTBT-Hex on a substrate with 20 transistors and its chemical structure. **b–e** Additional metalloporphyrin receptor molecules deposition and the corresponding chemical structures (TiO-TPP, Cu-TPP, and Zn-TPP). (f) Device principal scheme. **g** Device component layout with analog part with an array of operational amplifiers shown in red, drain/gate voltage controlling operational amplifiers in white, DC–DC converters in green, digital part with a microcontroller and microSD-card slot in blue and USB charger/data transfer port and power management ICs in orange. **h** Photo of the device with sensor chamber with pogo pins array and reference T/RH sensor from the top. Reprinted with permission from reference [96]. Copyright (2021) Springer Nature

functions. In another way, a chemically sensitive layer can connect to a commercial MOSFET to separate wet chemical work (sensing layer) to the transducing counterpart—the FET fabrication itself. Another essential aspect is the FET integration with electrical circuits or systems, for example, in smartphone-based devices to real-time response and remote data access aiming telemedicine [97]. In this regard, a challenge is developing a fully automated sensing platform for the simultaneous detection of several substances in real time and optimized for real applications (i.e., real world).

Among all FET types discussed in this chapter, OFETs are a good choice for biomedical devices fabrication due to their relative easy fabrication, semiconducting properties on the molecular level, the possibility of work with a wide range of organic compounds (conjugated polymers and small molecules), biocompatibility and flexibility. The chemical synthesis of organic compounds provides a wide variety of materials to be applied in OFETs. Among the OFETs, biocompatible FET platforms for implantable devices, flat transistors, and bio-integrated wearable systems are promising candidates for fully integrated systems with the patient body in continuous monitoring. On the other hand, advances in nanoscience and nanofabrication brought up the development of the NanoFETs, which paved the way for a new generation of devices for biomedical applications. However, reducing the length of the channel can bring some inherent problems like stability over operating mode and device performance at low power operation.

Although many efforts have been undertaken to achieve high-performance platforms with repeatability, better Figures of Merit, low cost, reusability, portability, and good stability with controlled threshold voltages, there are still particular challenges to be overcome. The true challenge is the FET large-scale production to commercial availability. Thus, FET-based devices still need efforts to leave the research centers to commercialization in real applications. Now, only a few companies are offering FET devices for biomedical applications. Examples are FET platform for the analysis of biofluid, in particular, sweat developed by Epicore Biosystems and commercialized by Gatorade and PepsiCo (https://www.epicorebiosystems.com/) and the label-free and real-time GFET sensor (Agile) to detect small molecules and proteins developed by Nanomed (https://nanomedical.com/) and explored with gFETs by the Cardea Bio (https://cardeabio.com/). These commercial advances indicate that FET transducers could indeed be applied in the biomedical field in the not-so-long future.

Acknowledgements The authors are grateful for the financial support provided by São Paulo Research Foundation (FAPESP: 18/07508-3).

References

1. Cao H, Coleman T, Hsiai TK, Khademhosseini A (2020) Interfacing bioelectronics and biomedical sensing. Springer, Berlin
2. Zhang A, Lieber CM (2016) Nano-bioelectronics. Chem Rev 116:215–257
3. Clark LC Jr, Lyons C (1962) Electrode systems for continuous monitoring in cardiovascular surgery. Ann N Y Acad Sci 102:29–45
4. Guilbault GG, Smith RK, Montalvo JG (1969) Use of ion selective electrodes in enzymic analysis. Cation electrodes for deaminase enzyme systems. Anal Chem 41:600–605
5. Peña-Bahamonde J, Nguyen HN, Fanourakis SK, Rodrigues DF (2018) Recent advances in graphene-based biosensor technology with applications in life sciences. J. Nanobiotechnology. 16:1–17
6. Janata J, Moss SD (1976) Chemically sensitive field-effect transistors. Biomed Eng (NY) 11:241–245
7. Caras S, Janata J (1980) Field effect transistor sensitive to penicillin. Anal Chem 52:1935–1937
8. Dzyadevych SV, Soldatkin AP, Anna V, Martelet C, Jaffrezic-Renault N (2006) Enzyme biosensors based on ion-selective field-effect transistors. Anal Chim Acta 568:248–258
9. Nguyen HH, Lee SH, Lee UJ, Fermin CD, Kim M (2019) Immobilized enzymes in biosensor applications. Mater (Basel) 12:121
10. Freire RS, Pessoa CA, Mello LD, Kubota LT (2003) Direct electron transfer: an approach for electrochemical biosensors with higher selectivity and sensitivity. J Braz Chem Soc 14:230–243

11. Huang Y, Sudibya HG, Fu D, Xue R, Dong X, Li L-J, Chen P (2009) Label-free detection of ATP release from living astrocytes with high temporal resolution using carbon nanotube network. Biosens Bioelectron 24:2716–2720
12. Park JW, Park SJ, Kwon OS, Lee C, Jang J (2014) Polypyrrole nanotube embedded reduced graphene oxide transducer for field-effect transistor-type H_2O_2 biosensor. Anal Chem 86:1822–1828
13. Minami T, Sato T, Minamiki T, Tokito S (2015) An extended-gate type organic FET based biosensor for detecting biogenic amines in aqueous solution. Anal Sci 31:721–724
14. Karunakaran C, Pandiaraj M, Santharaman P (2015) Immunosensors. In: Biosensors and bioelectronics. Elsevier, pp 205–245
15. Lipman NS, Jackson LR, Trudel LJ, Weis-Garcia F (2005) Monoclonal versus polyclonal antibodies: distinguishing characteristics, applications, and information resources. ILAR J 46:258–268
16. Matsumoto A, Miyahara Y (2013) Current and emerging challenges of field effect transistor based bio-sensing. Nanoscale 5:10702–10718. https://doi.org/10.1039/c3nr02703a
17. Schasfoort RBM, Bergveld P, Kooyman RPH, Greve J (1990) Possibilities and limitations of direct detection of protein charges by means of an immunological field-effect transistor. Anal Chim Acta. https://doi.org/10.1016/S0003-2670(00)80554-1
18. Halima HB, Bellagambi FG, Alcacer A, Pfeiffer N, Heuberger A, Hangouët M, Zine N, Bausells J, Elaissari A, Errachid A (2021) A silicon nitride ISFET based immunosensor for tumor necrosis factor-alpha detection in saliva. A promising tool for heart failure monitoring. Anal Chim Acta 1161:338468
19. Vieira NCS, Figueiredo A, Dos Santos JF, Aoki SM, Guimarães FEG, Zucolotto V (2014) Label-free electrical recognition of a dengue virus protein using the SEGFET simplified measurement system. Anal Methods 6. https://doi.org/10.1039/c4ay01803f
20. Magliulo M, De Tullio D, Vikholm-Lundin I, Albers WM, Munter T, Manoli K, Palazzo G, Torsi L (2016) Label-free C-reactive protein electronic detection with an electrolyte-gated organic field-effect transistor-based immunosensor. Anal Bioanal Chem 408:3943–3952. https://doi.org/10.1007/s00216-016-9502-3
21. Seshadri P, Manoli K, Schneiderhan-Marra N, Anthes U, Wierzchowiec P, Bonrad K, Di Franco C, Torsi L (2018) Low-picomolar, label-free procalcitonin analytical detection with an electrolyte-gated organic field-effect transistor based electronic immunosensor. Biosens Bioelectron 104:113–119
22. Minamiki T, Minami T, Kurita R, Niwa O, Wakida S, Fukuda K, Kumaki D, Tokito S (2014) A label-free immunosensor for IgG based on an extended-gate type organic field effect transistor. Materials (Basel) 7:6843–6852
23. Wang J, Ye D, Meng Q, Di C, Zhu D (2020) Advances in organic transistor-based biosensors. Adv Mater Technol 5:2000218
24. Kim K, Park C, Kwon D, Kim D, Meyyappan M, Jeon S, Lee J-S (2016) Silicon nanowire biosensors for detection of cardiac troponin I (cTnI) with high sensitivity. Biosens Bioelectron 77:695–701
25. Kong T, Su R, Zhang B, Zhang Q, Cheng G (2012) CMOS-compatible, label-free silicon-nanowire biosensors to detect cardiac troponin I for acute myocardial infarction diagnosis. Biosens Bioelectron 34:267–272
26. Ivanov YD, Romanova TS, Malsagova KA, Pleshakova TO, Archakov AI (2021) Use of silicon nanowire sensors for early cancer diagnosis. Molecules 26:3734
27. Hagen J, Lyon W, Chushak Y, Tomczak M, Naik R, Stone M, Kelley-Loughnane N (2013) Detection of orexin a neuropeptide in biological fluids using a zinc oxide field effect transistor. ACS Chem Neurosci 4:444–453. https://doi.org/10.1021/cn300159e
28. Gao A, Lu N, Wang Y, Li T (2016a) Robust ultrasensitive tunneling-FET biosensor for point-of-care diagnostics. Sci Rep 6:1–9
29. dos Santos FA, Vieira NCS, Zambianco NA, Janegitz BC, Zucolotto V (2020) The layer-by-layer assembly of reduced graphene oxide films and their application as solution-gated field-effect transistors. Appl Surf Sci 148698

30. Seo G, Lee G, Kim MJ, Baek SH, Choi M, Ku KB, Lee CS, Jun S, Park D, Kim HG, Kim SJ, Lee JO, Kim BT, Park EC, Kim S (2020) Il: rapid detection of COVID-19 causative virus (SARS-CoV-2) in human nasopharyngeal swab specimens using field-effect transistor-based biosensor. ACS Nano 14:5135–5142. https://doi.org/10.1021/acsnano.0c02823
31. Sheridan C (2020) COVID-19 spurs wave of innovative diagnostics. Nat Biotechnol 38:769–773
32. Shao W, Shurin MR, Wheeler SE, He X, Star A (2021) Rapid detection of SARS-CoV-2 antigens using high-purity semiconducting single-walled carbon nanotube-based field-effect transistors. ACS Appl Mater Interfaces 13:10321–10327. https://doi.org/10.1021/acsami.0c22589
33. Higuchi R, Fockler C, Dollinger G, Watson R (1993) Real time monitoring of DNA amplification reactions
34. Pachauri V, Ingebrandt S (2016) Biologically sensitive field-effect transistors: from ISFETs to NanoFETs. Essays Biochem 60:81–90
35. Rothberg JM, Hinz W, Rearick TM, Schultz J, Mileski W, Davey M, Leamon JH, Johnson K, Milgrew MJ, Edwards M (2011) An integrated semiconductor device enabling non-optical genome sequencing. Nature 475:348–352
36. Garrido-Cardenas JA, Manzano-Agugliaro F (2017) The metagenomics worldwide research. Curr Genet 63:819–829
37. Toumazou C, Shepherd LM, Reed SC, Chen GI, Patel A, Garner DM, Wang C-JA, Ou C-P, Amin-Desai K, Athanasiou P (2013) Simultaneous DNA amplification and detection using a pH-sensing semiconductor system. Nat Methods 10:641–646
38. Liu J, Chen X, Wang Q, Xiao M, Zhong D, Sun W, Zhang G, Zhang Z (2019) Ultrasensitive monolayer MoS_2 field-effect transistor based DNA sensors for screening of down syndrome. Nano Lett 19:1437–1444
39. Shariati M, Vaezjalali M, Sadeghi M (2021) Ultrasensitive and easily reproducible biosensor based on novel doped MoS_2 nanowires field-effect transistor in label-free approach for detection of hepatitis B virus in blood serum. Anal Chim Acta 1156:338360
40. Han D, Chand R, Kim Y-S (2017) Microscale loop-mediated isothermal amplification of viral DNA with real-time monitoring on solution-gated graphene FET microchip. Biosens Bioelectron 93:220–225
41. Hwang MT, Heiranian M, Kim Y, You S, Leem J, Taqieddin A, Faramarzi V, Jing Y, Park I, van der Zande AM (2020) Ultrasensitive detection of nucleic acids using deformed graphene channel field effect biosensors. Nat Commun 11:1–11
42. Li J, Wu D, Yu Y, Li T, Li K, Xiao M-M, Li Y, Zhang Z-Y, Zhang G-J (2021) Rapid and unamplified identification of COVID-19 with morpholino-modified graphene field-effect transistor nanosensor. Biosens Bioelectron 183:113206
43. Xu S, Jiang S, Zhang C, Yue W, Zou Y, Wang G, Liu H, Zhang X, Li M, Zhu Z (2018) Ultrasensitive label-free detection of DNA hybridization by sapphire-based graphene field-effect transistor biosensor. Appl Surf Sci 427:1114–1119
44. Huang C-C, Kuo Y-H, Chen Y-S, Huang P-C, Lee G-B (2021) A miniaturized, DNA-FET biosensor-based microfluidic system for quantification of two breast cancer biomarkers. Microfluid Nanofluidics 25:1–12
45. Priyanka R, Chandrasekar L, Shaik RR, Pradhan KP (2020) Label free DNA detection techniques using dielectric modulated FET: Inversion or tunneling? IEEE Sens J 21:2316–2323
46. Jinek M, Chylinski K, Fonfara I, Hauer M, Doudna JA, Charpentier E (2012) A programmable dual-RNA–guided DNA endonuclease in adaptive bacterial immunity. Science (80-.) 337:816–821
47. Pickar-Oliver A, Gersbach CA (2019) The next generation of CRISPR–Cas technologies and applications. Nat Rev Mol cell Biol 20:490–507
48. Li Y, Li S, Wang J, Liu G (2019) CRISPR/Cas systems towards next-generation biosensing. Trends Biotechnol 37:730–743
49. Hajian R, Balderston S, Tran T, DeBoer T, Etienne J, Sandhu M, Wauford NA, Chung J-Y, Nokes J, Athaiya M (2019) Detection of unamplified target genes via CRISPR–Cas9 immobilized on a graphene field-effect transistor. Nat Biomed Eng 3:427–437

50. Goldsmith BR, Locascio L, Gao Y, Lerner M, Walker A, Lerner J, Kyaw J, Shue A, Afsahi S, Pan D (2019) Digital biosensing by foundry-fabricated graphene sensors. Sci Rep 9:1–10
51. Balderston S, Taulbee JJ, Celaya E, Fung K, Jiao A, Smith K, Hajian R, Gasiunas G, Kutanovas S, Kim D (2021) Discrimination of single-point mutations in unamplified genomic DNA via Cas9 immobilized on a graphene field-effect transistor. Nat Biomed Eng 1–13
52. Grahame DC (1947) The electrical double layer and the theory of electrocapillarity. Chem Rev 41:441–501
53. Shoorideh K, Chui CO (2014) On the origin of enhanced sensitivity in nanoscale FET-based biosensors. Proc Natl Acad Sci 111:5111 LP – 5116 (2014). https://doi.org/10.1073/pnas.1315485111
54. Braus M (2019) The theory of electrolytes. I. Freezing point depression and related phenomena (Translation)
55. Stern E, Wagner R, Sigworth FJ, Breaker R, Fahmy TM, Reed MA (2007) Importance of the Debye screening length on nanowire field effect transistor sensors. Nano Lett 7:3405–3409
56. Wustoni S, Hideshima S, Kuroiwa S, Nakanishi T, Hashimoto M, Mori Y, Osaka T (2015) Sensitive electrical detection of human prion proteins using field effect transistor biosensor with dual-ligand binding amplification. Biosens Bioelectron 67:256–262. https://doi.org/10.1016/j.bios.2014.08.028
57. Baraban L, Ibarlucea B, Baek E, Cuniberti G (2019) Hybrid silicon nanowire devices and their functional diversity. Adv Sci 6:1900522
58. Huang Y-J, Lin C-C, Huang J-C, Hsieh C-H, Wen C-H, Chen T-T, Jeng L-S, Yang C-K, Yang J-H, Tsui F (2015) High performance dual-gate ISFET with non-ideal effect reduction schemes in a SOI-CMOS bioelectrical SoC. In: 2015 IEEE international electron devices meeting (IEDM). IEEE, pp 22–29
59. Elnathan R, Kwiat M, Pevzner A, Engel Y, Burstein L, Khatchtourints A, Lichtenstein A, Kantaev R, Patolsky F (2012) Biorecognition layer engineering: overcoming screening limitations of nanowire-based FET devices. Nano Lett 12:5245–5254
60. Jang H-J, Ahn J, Kim M-G, Shin Y-B, Jeun M, Cho W-J, Lee KH (2015) Electrical signaling of enzyme-linked immunosorbent assays with an ion-sensitive field-effect transistor. Biosens Bioelectron 64:318–323. https://doi.org/10.1016/j.bios.2014.09.020
61. Gao N, Gao T, Yang X, Dai X, Zhou W, Zhang A, Lieber CM (2016b) Specific detection of biomolecules in physiological solutions using graphene transistor biosensors. Proc Natl Acad Sci 113:14633–14638
62. Kulkarni GS, Zhong Z (2012) Detection beyond the debye screening length in a high-frequency nanoelectronic biosensor. Nano Lett 12:719–723. https://doi.org/10.1021/nl203666a
63. Bergveld P (1972) Development, operation, and application of the ion-sensitive field-effect transistor as a tool for electrophysiology. IEEE Trans Biomed Eng 342–351
64. Bergveld P, Wiersma J, Meertens H (1976) Extracellular potential recordings by means of a field effect transistor without gate metal, called OSFET. IEEE Trans Biomed Eng 136–144
65. Fromherz P, Offenhausser A, Vetter T, Weis J (1991) A neuron-silicon junction: a Retzius cell of the leech on an insulated-gate field-effect transistor. Science (80-.). 252:1290–1293
66. Krause M, Ingebrandt S, Richter D, Denyer M, Scholl M, Sprössler C, Offenhäusser A (2000) Extended gate electrode arrays for extracellular signal recordings. Sens Actuators B Chem 70:101–107
67. Ecken H, Ingebrandt S, Krause M, Richter D, Hara M, Offenhäusser A (2003) 64-Channel extended gate electrode arrays for extracellular signal recording. Electrochim Acta 48:3355–3362
68. Chang K-S, Sun C-J, Chiang P-L, Chou A-C, Lin M-C, Liang C, Hung H-H, Yeh Y-H, Chen C-D, Pan C-Y (2012) Monitoring extracellular K+ flux with a valinomycin-coated silicon nanowire field-effect transistor. Biosens Bioelectron 31:137–143
69. Nutt JG, Carter JH, Sexton GJ (2004) The dopamine transporter: importance in Parkinson's disease. Ann Neurol Off J Am Neurol Assoc Child Neurol Soc 55:766–773
70. Li B-R, Hsieh Y-J, Chen Y-X, Chung Y-T, Pan C-Y, Chen Y-T (2013) An ultrasensitive nanowire-transistor biosensor for detecting dopamine release from living PC12 cells under hypoxic stimulation. J Am Chem Soc 135:16034–16037

71. Park SJ, Lee J, Seo SE, Kim KH, Park CS, Lee SH, Ban HS, Lee BD, Song HS, Kim J (2020) High-performance conducting polymer nanotube-based liquid-ion gated field-effect transistor aptasensor for dopamine exocytosis. Sci Rep 10:1–12
72. Kireev D, Zadorozhnyi I, Qiu T, Sarik D, Brings F, Wu T, Seyock S, Maybeck V, Lottner M, Blaschke BM (2016) Graphene field-effect transistors for in vitro and ex vivo recordings. IEEE Trans Nanotechnol 16:140–147
73. Kyndiah A, Leonardi F, Tarantino C, Cramer T, Millan-Solsona R, Garreta E, Montserrat N, Mas-Torrent M, Gomila G (2020) Bioelectronic recordings of cardiomyocytes with accumulation mode electrolyte gated organic field effect transistors. Biosens Bioelectron 150:111844
74. Li W-C, Soffe SR, Roberts A (2004) A direct comparison of whole cell patch and sharp electrodes by simultaneous recording from single spinal neurons in frog tadpoles. J Neurophysiol
75. Duan X, Gao R, Xie P, Cohen-Karni T, Qing Q, Choe HS, Tian B, Jiang X, Lieber CM (2012) Intracellular recordings of action potentials by an extracellular nanoscale field-effect transistor. Nat Nanotechnol 7:174–179
76. Gao R, Strehle S, Tian B, Cohen-Karni T, Xie P, Duan X, Qing Q, Lieber CM (2012) Outside looking in: nanotube transistor intracellular sensors. Nano Lett 12:3329–3333
77. Zhao Y, You SS, Zhang A, Lee J-H, Huang J, Lieber CM (2019) Scalable ultrasmall three-dimensional nanowire transistor probes for intracellular recording. Nat Nanotechnol 14:783–790
78. Jobling DT, Smith JG, Wheal H (1981) V: Active microelectrode array to record from the mammalian central nervous system in vitro. Med Biol Eng Comput 19:553–560
79. Sprössler C, Denyer M, Britland S, Knoll W, Offenhäusser A (1999) Electrical recordings from rat cardiac muscle cells using field-effect transistors. Phys Rev E 60:2171
80. Spanu A, Martines L, Bonfiglio A (2021) Interfacing cells with organic transistors: a review of in vitro and in vivo applications. Lab Chip 21:795–820
81. Cramer T, Chelli B, Murgia M, Barbalinardo M, Bystrenova E, de Leeuw DM, Biscarini F (2013) Organic ultra-thin film transistors with a liquid gate for extracellular stimulation and recording of electric activity of stem cell-derived neuronal networks. Phys Chem Chem Phys 15:3897–3905
82. Desbief S, Di Lauro M, Casalini S, Guerin D, Tortorella S, Barbalinardo M, Kyndiah A, Murgia M, Cramer T, Biscarini F (2016) Electrolyte-gated organic synapse transistor interfaced with neurons. Org Electron 38:21–28
83. Qing Q, Pal SK, Tian B, Duan X, Timko BP, Cohen-Karni T, Murthy VN, Lieber CM (2010) Nanowire transistor arrays for mapping neural circuits in acute brain slices. Proc Natl Acad Sci 107:1882–1887
84. Timko BP, Cohen-Karni T, Yu G, Qing Q, Tian B, Lieber CM (2009) Electrical recording from hearts with flexible nanowire device arrays. Nano Lett 9:914–918
85. Cheng Z, Hou J, Zhou Q, Li T, Li H, Yang L, Jiang K, Wang C, Li Y, Fang Y (2013) Sensitivity limits and scaling of bioelectronic graphene transducers. Nano Lett 13:2902–2907
86. Masvidal-Codina E, Illa X, Dasilva M, Calia AB, Dragojević T, Vidal-Rosas EE, Prats-Alfonso E, Martínez-Aguilar J, Jose M, Garcia-Cortadella R (2019) High-resolution mapping of infraslow cortical brain activity enabled by graphene microtransistors. Nat Mater 18:280–288
87. Hébert C, Masvidal-Codina E, Suarez-Perez A, Calia AB, Piret G, Garcia-Cortadella R, Illa X, Del Corro Garcia E, De la Cruz Sanchez JM, Casals DV (2018) Flexible graphene solution-gated field-effect transistors: efficient transducers for micro-electrocorticography. Adv Funct Mater 28:1703976
88. Yogeswaran N, Hosseini ES, Dahiya R (2020) Graphene based low voltage field effect transistor coupled with biodegradable piezoelectric material based dynamic pressure sensor. ACS Appl Mater Interfaces 12:54035–54040
89. Kim Y, Chortos A, Xu W, Liu Y, Oh JY, Son D, Kang J, Foudeh AM, Zhu C, Lee Y (2018) A bioinspired flexible organic artificial afferent nerve. Science (80-.) 360:998–1003

90. Feili D, Schuettler M, Doerge T, Kammer S, Stieglitz T (2005) Encapsulation of organic field effect transistors for flexible biomedical microimplants. Sens Actuators Phys 120:101–109
91. Gerasimov JY, Gabrielsson R, Forchheimer R, Stavrinidou E, Simon DT, Berggren M, Fabiano S (2019) An evolvable organic electrochemical transistor for neuromorphic applications. Adv Sci 6:1801339
92. Zoumpoulidis T, Prodromakisy T, Van Zeijl H, Michelakisy K, Bartek M, Toumazouy C, Dekkerz R (2009) Stretchable array of ISFET devices for applications in biomedical imagers. In: Sensors, 2009 IEEE. IEEE, pp 7–12
93. Zeng J, Miscourides N, Georgiou P (2018) A 128× 128 current-mode ultra-high frame rate ISFET array for ion imaging. In: 2018 IEEE international symposium on circuits and systems (ISCAS). IEEE, pp 1–5
94. Zeidell AM, Ren T, Filston DS, Iqbal HF, Holland E, Bourland JD, Anthony JE, Jurchescu OD (2020) Organic field-effect transistors as flexible, tissue-equivalent radiation dosimeters in medical applications. Adv Sci 7:2001522
95. Tang X-W, Yang Y, Kim W, Wang Q, Qi P, Dai H, Xing L (2005) Measurement of ionizing radiation using carbon nanotube field effect transistor. Phys Med Biol 50:N23
96. Anisimov DS, Chekusova VP, Trul AA, Abramov AA, Borshchev OV, Agina EV, Ponomarenko SA (2021) Fully integrated ultra-sensitive electronic nose based on organic field-effect transistors. Sci Rep 11:1–12
97. Banik S, Melanthota SK, Vaz JM, Kadambalithaya VM, Hussain I, Dutta S, Mazumder N (2021) Recent trends in smartphone-based detection for biomedical applications: a review. Anal Bioanal Chem 1–18

Fundamentals for Virus and Antigen Detection in Immunotechnologies

Karla Ribeiro Castro, Sthéfane Valle de Almeida, Ronaldo Censi Faria, and Frank N. Crespilho

Abstract Viruses are agents responsible for many human infectious diseases, including influenza, HIV, and the novel coronavirus. Despite efforts, the completed spectrum of viruses that can infect human host cells as well their impact on health remains unknown. Nowadays, there are several technologies for the detection of viruses and its antigens, although it is still necessary analytical improvements, the use of appropriated biomarkers remains crucial for a reliable detection. The main method for viral detection is the polymerase chain reaction (PCR)-based assays. However, difficulties such as time-consuming analysis, multiple-step assay, machinery, and reagents can be a challenge in the diagnosis. In this chapter, we briefly discuss biomarkers that are normally used for virus and antigen detections in different sensor platforms.

Keywords Viral biomarkers · Biosensors · COVID-19 · Proteins · Immunosensor

1 Introduction

IUPAC (International Union of Pure and Applied Chemistry) defined biosensors as *"A device that uses specific biochemical reactions mediated by isolated enzymes, immunosystems, tissues, organelles or whole cells to detect chemical compounds usually by electrical, thermal or optical signals"* [56]. In general, the biosensor format consists in the presence of a bioreceptor, transducer, electronics and display, for more information, we recommend a specialized literature [3, 7]. When the first glucose oxidase sensor aiming to detect blood glucose was developed by Clark and Lyons 60 years ago, a tremendous increase of technologies related to biosensing phenomenon was observed [18, 24]. Since then, many biosensors and methodologies

K. R. Castro · F. N. Crespilho (✉)
Institute of Chemistry of São Carlos (IQSC), University of São Paulo (USP), São Carlos, SP 13560-970, Brazil
e-mail: frankcrespilho@iqsc.usp.br

S. V. de Almeida · R. C. Faria
Department of Chemistry, Federal University of São Carlos, São Carlos, SP 13565-905, Brazil

F. N. Crespilho (ed.), *Advances in Bioelectrochemistry Volume 3*,
https://doi.org/10.1007/978-3-030-97921-8_2

have been developed for various applications, ranging from molecules detection to quantification of medical analytes [12, 23, 33, 58, 78, 98, 108, 112, 123]. Despite the large variety of techniques that biosensors can be used, this chapter will be focused on, especially in the detection of viruses and their biomarkers.

Viruses are defined as an infectious unit (diameters of about 16 to 300 nm) and they present a diverse mechanism of action. They contain a DNA or RNA genome surrounded by capsid proteins, some of which are used to gain entry to the host cell [18, 83]. Viruses are responsible for numerous diseases worldwide, leading to many deaths annually. Although they may diverge in terms of the diseases they cause, all viruses require a host to propagate and survive [83, 119]. Most of them (common cold, for instance) can be eradicated from the human body through the immune response. However, more complicated viruses might not be well controlled, even threatening host survival.

In the pandemic years caused by the severe acute respiratory syndrome coronavirus 2 (SARS-CoV-2), it is clear the importance of the detection and quantification of virus and their biomarkers, especially for diagnosis and therapeutics. Biomarkers refer to anything that can be used as an indicator of a particular disease state or some other physiological state of an organism, that reflects the severity or presence of some disease state which can be measured accurately and reproducibly [52, 115].

It is well-known that, normally, viral infections may cause generic symptoms, making it difficult to use only the symptoms for a precise diagnostic. Their detection requires some specific equipment, methodologies and techniques such as RT-PCR—reverse transcription polymerase chain reaction (nucleic acid detection), ELISA—Enzyme-Linked Immunosorbent Assay (antibody or antigen detection) and gene sequencing [84]. These diagnostic tools are essential for identifying the disease-causing virus. Currently, new diagnostic methods are being developed for viral diseases, most involving the use of biosensors for biomarkers detection.

So, what kind of biomarkers could be used for viral disease detection? To answer this question, it is important to choose which viral disease is the subject and considerate that the biochemical mechanisms can be different in each infection, as well as the biomarkers. In this sense, each biomarker reveals information about the viral infection, whether about the virus presence or the infection window. Besides, for each biomarker, there is an appropriate type of sample (blood, saliva, etc.). In this way, it is possible not only to identify the virus, but also to understand at what stage the infection is, according to the biomarker found in each sample.

Biosensors could help doctors in clinical diagnoses, assisting them to identify the specific virus, and, if is the case, aiding to the prescription of an appropriate treatment. In addition, the presence of selective/specific antibodies can be detected employing viral components as sensing components to demonstrate a history of infection and the specific antibodies can be used as sensing agents to detect viral components [18, 34].

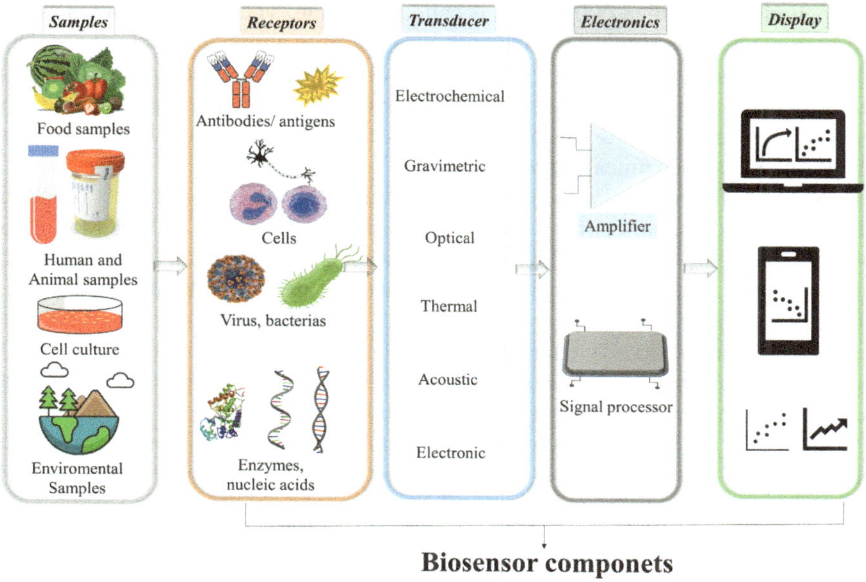

Fig. 1 Schematic representation of a biosensor [46, 94]

2 Biosensors and Immunoassays Designs

As well-known, biosensors are receptor-transducer integrated analytical devices, in which the receptor or biorecognition element is responsible for the specificity, and the transducer converts the recognition event into a measurable signal [88]. Normally, they are categorized according to the type of receptor element involved, listed in Fig. 1 [94].

Biosensors that are based on the immunological interaction between antigen and antibody are called immunosensors. These can use transducers of various types, among which we highlight the optical and electrochemical. Immunosensors have greater potential for use as point-of-care, as they are generally stable and generate quick and accurate responses [59].

3 Proteins in Viral Detection

3.1 Proteins

Briefly, proteins are organic molecules of high molecular weight (greater than 6000 Da) formed by two or more amino acids through a peptide bond. When the compound has less than 6000 Da, it is called a peptide. There are 20 known amino acids in nature, and this leads to a variety of possible amino acid sequences. Due to this, there is a great plurality of existing protein structures and each one plays a fundamental role in nature [8, 95]. For each biological function, there is a type of protein involved. Proteins can be classified according to their function, molecular shape and physicochemical properties [8].

The three-dimensional structure of the protein is directly related to its function, since a specific spatial conformation is necessary for it to bind and interact in the organism [28, 95]. There are eleven types of proteins based on their biological function: catalytic proteins; transport protein; structural or cytoskeletal proteins (e.g., collagen); hormonal or effector proteins (e.g., insulin); electron transfer proteins (e.g., cytochrome oxidase); chaperones (e.g., DnaK); repressor proteins (e.g., Cro); contractile proteins (e.g., actin); genetic proteins (e.g., nucleoproteins); receptors (e.g., CD4); and protective proteins (e.g., antibodies) [129, 136].

Antibodies are the main element in immunosensing technologies. Furthermore, other types of protein also play a key role, acting as biomarkers of infections. A biomarker can be any parameter that indicates normal or biological processes, such as the change in the activity of a substance produced by the organism or the presence of proteins of a virus. We will see more potential virus biomarkers later.

To understand the role of antibodies, it is first necessary to know the basics of how the immune system works.

3.2 Immune System and the Role of Antibodies

The immune system is made up of different proteins. Its function is to protect the body from any agent that can cause infection. This includes the presence of external agents (viruses, bacteria, fungi, parasites) and dysfunctions that can spontaneously occur in the body, such as cancer cells [90, 138].

In most cases, innate immunity can eliminate threats. In vertebrates, when invaders persist, the adaptive immune system is activated. Every biomolecule that can activate the immune system is called an antigen. When identified, the antigen is recognized and attacked according to its characteristics. Thereby, a specific immune response to the first infection takes days, but the response to future exposures occurs faster [25, 48]. Adaptive immunity involves the production of T and B lymphocytes. As a result, B lymphocytes produce antibodies, also known as immunoglobulins (Ig).

Antibodies have the function of neutralizing the pathogen, preventing it from multiplying. On the first contact with the antigen, the B cell produces specific antibodies to neutralize it. This contact can occur through exposure to the pathogen or vaccination [25, 49]. They are formed by two light chains (~23 kDa) and two heavy chains (~50 kDa) of polypeptides. The polypeptide chains are linked together by disulfide bridges and form a structure similar to the letter *Y*. These chains contain a constant region that is similar in all antibodies of the same class and a variable region

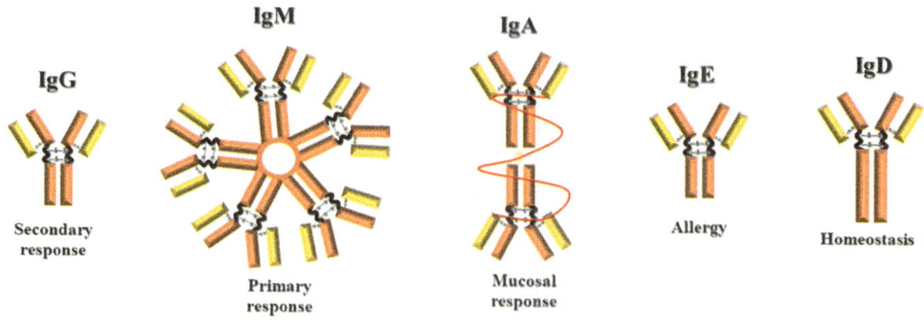

Fig. 2 Schematic representation of the classes of human antibodies (IgM, IgG, IgA, IgD, and IgE) and their estimated concentration in serum [11, 107, 130]

that is intended for binding with the specific antigen, so it is different for each target. In this region, every antibody molecule can bind to two antigens at a time [25, 49].

There are five classes of human antibodies (IgM, IgG, IgA, IgD, and IgE) and each class has its function determined by the constant region of the heavy chains, that differ from each other in structure and size [11, 13]. Different antibodies (Fig. 2) can be detected at different stages of an infection. In this sense and due to their high specificity, antibodies can be used to detect the presence of many pathogens. We will see more of these proteins and how they can be used for virus detecting.

3.3 Antibodies Used in the Diagnosis of Viral Infection

Antibodies have been used in different diagnostic strategies. With high specificity and the ability to detect low concentrations, they can be employed in immunosensors and immunoassays to detect diverse viruses. For immunodiagnostics, antibodies are always related to a protein whose abnormal concentration or presence in the body indicates a viral infection [16]. These proteins are called biomarkers. Any protein whose activity can be related to the presence of a virus in the organism can be considered a biomarker. They can be used to indicate infections or to monitor their progression. Therefore, many biomarkers contribute to the diagnosis and prognosis of virus-related diseases [16].

The use of an isolated biomarker to detect its specific antibody in biological fluids is common. It should be noted that the detection of specific antibodies does not necessarily indicate the presence of the active virus. However, it can indicate if there was a previous infection and assess the degree and duration of the immune response, which is important for analyzing real data on infection and the effectiveness of vaccines and therapeutic actions [71, 102]. Furthermore, detection of antibodies in immunocompromised patients can be difficult as antibody production can be insufficient to be detected. In these cases, it is not possible to assess whether the infection is in the acute phase or not [100, 142]. Thereby, tests for the detection of antibodies must be evaluated very carefully.

In theory, any class of antibodies can identify a virus infection (present or past). However, for it to be used as a diagnostic tool, the antibody must be found in the organism in a concentration that can be measured with adequate precision and accuracy. For example, IgD and IgE antibodies match less than 1% of total antibodies in serum. Thus, detection of these antibodies is difficult and not very sensitive. Besides, they have a half-life of fewer than 3 days, which indicates that the tests developed with these antibodies must be performed in a limited window of time [11]. In this context, the most applied antibodies in virus immunosensing strategies are IgA, IgM, and IgG due to their high concentration in serum samples and wide half-life in the body [11].

Characteristic of mucous membranes, IgA is a monomer (~160 kDa) and represents 15% of antibodies in serum (2.0 mg mL^{-1}). According to its structure, its half-life is 5–6 days and thus can be used to detect the presence of viruses in the first days of infection [113, 137]. Some studies report a greater specificity of IgA antibodies compared to IgM for the detection of flaviviruses, especially for secondary infections. Furthermore, as a marker of early infection, detection by IgA allows for a more accurate estimation of the date of infection [29, 134]. In addition, there are studies on the relationship between IgA concentration and the severity of COVID-19 cases [107, 139]. Although it is less used than IgM and IgG, it has recently been applied in commercial tests for virus diagnosis [31, 81].

IgM accounts for about 10% of the antibodies in the body and can be found in serum at a concentration of 1.2 mg mL^{-1}. It is the antibody with the highest molecular weight (~1000 kDa), as it is a pentamer composed of five subunits and can consequently bind to ten antigens. In addition to neutralizing the antigen, IgM is also responsible for activating the complement system. Once the virus enters the organism, it is the primary responsibility of the adaptive immune system as the first antibody to be produced. So, the presence of IgM in serum indicates that the infection is in its acute stage. With a half-life of 5 days, it generates a short-term immunity that decays as IgG begins to be produced [12, 107, 111]. However, specific IgM from certain viruses can be found in the body for longer. Studies report that Zika virus-specific IgM antibodies were found in the serum of patients up to 1 year after the onset of infection [114]. Similarly, it is believed that specific dengue IgM can be found up to 6 months after the onset of symptoms [22, 97]. Thus, the presence of IgM antibodies does not necessarily indicate acute infection and the diagnosis should not be based on this alone. Possible cases of reinfection, especially in epidemic areas, should also be carefully evaluated to avoid false-positive results.

In turn, IgG antibodies are the secondary response of the immune system. They are the most abundant antibodies in serum (75%), with a concentration of approximately 20 mg mL^{-1}. It is also the longest half-life (~23 days). It is considered the main agent against reinfections, as it continues to be synthesized by the body for years after systemic infections [6, 11]. In addition to neutralizing, IgG antibodies act in opsonization by binding to the antigen to facilitate phagocytosis and in the process of antibody-dependent cellular cytotoxicity (ADCC), binding to infected cells for elimination by NK cells [122, 133]. Although they do not indicate the acute phase of the infection, IgGs can be used to monitor contact with a specific virus, evaluating the seroprevalence in the population. This type of test is especially important for areas where there is or has been an epidemic. For example, in Cebu (Philippines), it was found that 89.3% of 2996 children between the ages of 8 and 14 had contact with the dengue virus [73]. In southern Mali (Guinea), it was possible to assess that 1.5%–6.1% of the 600 people surveyed were exposed to the Ebola virus [132].

Rapid tests are found to detect IgM/IgG, most of them colorimetric. Currently, there are IgM/IgG tests for diagnosing many viruses such as arboviruses as dengue [51], chikungunya [96], and Zika [74]), HIV[42], hepatitis E [132], and recently COVID-19 [4]. To date, however, there are no commercially available serological electrochemical tests.

3.4 Other Proteins as Antigens for Virus Immunodetection

Some biomolecules normally existing in the body can change their concentration according to the origin and intensity of infection. In particular, some proteins can differentiate between an infection caused by a virus or a bacteria, which is important to guide medical treatment [104]. These proteins are generally involved in the inflammatory and immune processes. Most of them respond to both viral and bacterial infections, although they have higher concentrations for bacteria, enabling differentiation [99]. Of these, we can highlight C-reactive protein (CRP) due to the high sensitivity of the electrochemical assays reported in the literature [47, 75, 120]. Although commonly used for bacterial detection, studies indicate that high CPR concentrations (often reaching levels of bacterial infections, >100 mg L^{-1}) [118] can potentially predict severe cases of disease caused by SARS-CoV-2 [70] and H1N1 [128]. Thus, CRP can also be applied in the prognosis of viral diseases.

Interleukins (IL), a type of cytokine, are also used as biomarkers to differentiate between different types of pathogens. Similar to CRP, ILs have higher concentrations in blood when the infection is caused by a bacterium than a virus [72, 99], and we can highlight that especially IL-2, IL-8 and IL-10 have the potential to distinguish the types of infections [140]. Recently, an aerosol-jet-printed graphene-based impedimetric immunosensor was developed and applied to the monitoring of IL-10. Anti-IL-10 antibodies were immobilized to the graphene surface, obtaining a detection limit of 46 pg mL^{-1} [93]. Although this LOD is below the concentration usually found in viral infections (~14.27 pg mL^{-1}), it is enough to detect patients with bacteria (~205.11 pg mL^{-1}), enabling them to identify the source of infection [72].

Protein-coding genes can also be used to make this diagnosis. As an example, Interferon Alpha Inducible Protein 27 (IFI27 protein) may be a biomarker to differentiate respiratory infections caused by the influenza virus from bacterial ones [45, 117]. In general, detections like this are performed by genetic methods such as polymerase chain reaction (PCR) and genosensors (biosensors that use genetic material).

In this sense, no immunosensors or immunoassays for the detection of IFI27 were found in the literature.

3.5 Viral Proteins as Biomarkers

Viral proteins can also be considered diagnostic biomarkers, as they can be detected to identify a virus presence. In addition to differentiating viral from bacterial infections, these proteins can identify which virus is causing it. They can be present in specific regions of a certain virus (capsid or envelope), or be produced in the infected cell [38]. Most immunoassays and sensors use viral proteins as antigenic targets. These devices have high specificity due to antigen–antibody binding and are capable of detecting the presence of a particular virus. Its sensitivity, although also high, is not comparable with nucleic acid amplification techniques such as PCR. However, immunoassays that detect viral proteins have faster responses and most systems are simpler to handle (especially electrochemical ones). Thus, immunodiagnostic methods with viral antigens are more accessible to detect active infections [17, 62].

Viral proteins can be used to diagnose different hepatitis viruses. They can be components of the capsid, such as the core protein from the hepatitis C virus (HCV) [55, 67]. Core protein can be found in the serum in the acute phase of the infection. It can be detected 1 or 2 days after the emergence of HCV RNA, which indicates viral replication [125]. Commercial tests for diagnosis of HCV use core protein detection and have high sensitivity and specificity [43]. The detection of this protein has shown high sensitivity and can be fast and inexpensive when combined with electrochemical techniques. In this context, electrochemical immunosensors have been developed, although few works can be found in the literature. As an example, we can mention a sandwich immunosensor based on Nafion@TiO_2 nanoparticles and Celestine Blue. This device was able to detect on the fg mL^{-1} range using differential-pulse voltammetry (DPV). Furthermore, it was considered stable for several days and also showed good precision and accuracy when applied to spiked serum samples [126].

Regarding the hepatitis B virus (HBV), there are two biomarkers. HBsAg (surface antigen) indicates acute-phase infection. It is an envelope protein and may indicate chronic infection if it persists for more than 6 months in the blood. Another biomarker, the HBeAg protein is produced by the virus in the viral replication phase. When detected in blood, it indicates that infectivity is high [30, 64]. Electrochemical immunosensors for the detection of these antigens usually present detection limits in the range of ng mL^{-1}, sometimes achieving fg mL^{-1}. These devices usually exhibit superior sensitivity to ELISA-based immunoassays [1, 61, 124].

The NS1 protein is secreted in large amounts by all flaviviruses. As it is the first non-structural protein to be translated, it can be detected one day after the onset of infection. Although NS1 proteins are similar in all flaviviruses, the use of specific antibodies is capable of making specific diagnoses for each infection. Therefore, it is often used as an antigen to identify which virus caused the viremia [86, 92].

To improve the diagnosis, generally, detection of NS1 and its specific antibodies is performed. As NS1 can be detected in serum in a short window of time, identifying the presence of IgM or IgG can increase the sensitivity and make it possible to recognize a secondary infection [105, 106]. In addition, studies have been carried out on the detection of the NS1 protein of the dengue virus in mosquitoes, aiming to help control transmission and prevent possible outbreaks [131, 135].

Electrochemical immunosensors for the detection of NS1, especially for the dengue virus, have been developed. A disposable portable system with a low detection limit and low cost can indicate an advance in dengue control and can be used on a large scale in endemic areas [5, 89]. Above all, a point-of-care device without flavivirus cross-reaction is essential. Although studies have already shown encouraging results regarding the selectivity of these immunodiagnostics [40, 60], this is still a challenge.

3.6 Viral Protein for Novel Coronavirus Detection (COVID-19)

Several reviews already related the impact of the development of medical devices, especially in the COVID-19 pandemic scenario [80], and the potentiality of electrochemical biosensors use for diagnoses [77]. For the diagnosis of human coronaviruses, especially SARS-CoV-2, viral proteins have been used in antigen detection tests. The most explored are the nucleocapsid (N) and spike (S) proteins. The N protein is found in the nucleus binding to viral RNA and is considered essential for viral replication and genome packaging. It is highly antigenic, so most IgG antibodies produced (whether by infection or vaccination) are predominantly directed to it [26, 79]. In turn, glycoprotein S is present on the surface of the virus. It mediates the virus's interaction with the host cell through a receptor, allowing its entry [50, 65]. Furthermore, it is the main protein targeted in vaccines for coronaviruses [27, 39]. IgGs are the most used to detect previous COVID-19 infection, and IgM is related to recent contamination. However, IgA can be used since they are responsible for the immune exclusion process. It has been illustrated that elevated total IgA and IgA-aPL (IgA-antiphospholipid) can be notably associated with severe COVID-19 infection [19, 91, 103]

Electrochemical immunoassays for detection of both these proteins of SARS-CoV-2 present highly sensitive results, with reduced analysis time (~10 min) compared to standard techniques [69]. However, the sensitivity of these methods still needs to be improved to allow the detection of asymptomatic or low viral load patients. Besides, like flaviviruses, coronaviruses also have similarities to each other, and this can lead to interference in the diagnosis. In particular, SARS-CoV and SARS-CoV-2 show high cross-reactivity, since there is a 90% similarity of amino acids of protein N in both viruses [66]. Although protein S also shows high similarity (77%), studies claim that antibodies are highly specific and thus there is no significant interference of different coronaviruses [57].

Lateral flow normally allows a qualitative and/or quantitative analysis and their use could assist nucleic acid tests, providing an extended window for detecting active and recent infections [20, 103]. It is already reported in the literature various lateral flow immunoassay for detection of IgM and IgG in blood [21, 68, 91]. Some works aim the comparation of different LFIA for COVID-19 detections [37, 41, 85, 121]. In general, LFIA presents a sensibility ranging from 37.7 to 99.2% and specificity from 92.4 to 100.0% [82].

In Table 1 it is presented an overall use of different antigens biomarkers for virus detection.

4 Final Considerations

Viruses are a threat to the health and well-being of humans, infecting host cells and causing many chronic and acute diseases. In addition to treatments in general being only symptomatic, detecting and identifying the virus causing an infection is still a challenge. In this sense, biosensors may be a viable option for the accurate and rapid detection of these endemic agents, especially those based on the formation of antigen–antibody complexes. These systems provide fast detection, low cost, and ease of handling without the need for training and centralized laboratories. Serological tests are already available on the market to detect several viruses, and can be used especially for clinical diagnoses, providing a set of benefits to the patient, the doctor, and the entire health system as it helps in disease early detection, treatment, and monitoring. Furthermore, they can be used to monitor endemic areas, helping to break the transmission chain quickly and thus prevent possible outbreaks.

Regardless of these advantages, immunotechnologies applied to clinical diagnoses still need to be improved. The COVID-19 crisis exposed the necessity of upgrading the diagnostic tools for virus detection with high accuracy, reliability, and speed. Commercial devices more sensitive to recent infections, capable of accurately quantifying biomarkers and selectively identifying the causative virus are still scarce, although studies are advancing in this area. There are many works in the literature reporting the development of devices employing different detection strategies with great potential to help identify viruses. In the next few years, these devices may come off the paper, thus meeting the need for accurate sensing platforms with high selectivity and sensitivity levels.

Table 1 Different methods used in viral detection

Method	Advantages	Disadvantages	Disease	Biomarker	Literature
ELISA	Different formats, simple assay design, amenable high volume-testing	Long runtime, sample matrix can be an interference, false negatives	COVID-19	S and N protein	[2]
			Zika	NS1 protein	[10]
			Ebola	soluble GP	[44]
			HIV	HIV-1 p24	[87]
LFIA	Rapid (10–30 min), user-friendly, easy to interpret	Semi-quantitative, not achieve comparable to nucleic-acid detection	COVID-19	S and N protein	[76]
			Ebola	GP and VP40	[35]
			HIV	p24 protein	[141]
			Dengue	NS1 protein	[63]
Electrochemical immunosensors (Antigen detection)	Low-cost, portable, wide range of applications, simplicity, miniaturization	Sensitive to sample matrix effects, stability,	COVID-19	S protein	[53]
			Zika	Envelop protein	[60]
			Dengue	NS1 protein	[5]
			Malaria	Ag-*Pf*HRP2	[36]
			HIV	gp20	[109]
			Hepatitis C	HCV	[127]
			Hepatitis B	HBsAg	[9]
Electrochemical immunosensors (Antibody detection)	–	–	COVID-19	Anti-SARS-CoV-2 IgG	[101]
			Zika Virus	Anti-Zika IgG	[14]
			Dengue	Anti-Denv IgG	[110]
			Measles virus	Anti-Measles	[116]
			West Nile virus	Anti-West Nile IgG	[54]
			Hepatitis C	anti-HBc IgG	[15]
			Hepatitis B	anti-hepatitis B IgG	[32]

References

1. Akkapinyo C, Khownarumit P, Waraho-Zhmayev D, Poo-arporn RP (2020) Development of a multiplex immunochromatographic strip test and ultrasensitive electrochemical immunosensor for hepatitis B virus screening. Anal Chim Acta 1095:162–171. https://doi.org/10.1016/j.aca.2019.10.016
2. Algaissi A, Alfaleh MA, Hala S et al (2020) SARS-CoV-2 S1 and N-based serological assays reveal rapid seroconversion and induction of specific antibody response in COVID-19 patients. Sci Rep 10. https://doi.org/10.1038/s41598-020-73491-5
3. Ali J, Najeeb J, Asim Ali M et al (2017) Biosensors: their fundamentals, designs, types and most recent impactful applications: a review. J Biosens Bioelectron 08:1–9. https://doi.org/10.4172/2155-6210.1000235
4. Andrey DO, Cohen P, Meyer B et al (2020) Head-to-head accuracy comparison of three commercial COVID-19 IgM/IgG serology rapid tests. J Clin Med 9:2369. https://doi.org/10.3390/jcm9082369
5. Awan M, Rauf S, Abbas A et al (2020) A sandwich electrochemical immunosensor based on antibody functionalized-silver nanoparticles (Ab-Ag NPs) for the detection of dengue biomarker protein NS1. J Mol Liq 317:114014. https://doi.org/10.1016/j.molliq.2020.114014
6. Beenhouwer DO (2018) Molecular basis of diseases of immunity, Second Edi. Elsevier Inc.
7. Bhalla N, Jolly P, Formisano N, Estrela P (2016) Introduction to biosensors. Essays Biochem 60:1–8. https://doi.org/10.1042/EBC20150001
8. Blanco A, Blanco G (2017) Proteins. In: Medical biochemistry. Elsevier, pp 21–71
9. Boonkaew S, Yakoh A, Chuaypen N et al (2021) An automated fast-flow/delayed paper-based platform for the simultaneous electrochemical detection of hepatitis B virus and hepatitis C virus core antigen. Biosens Bioelectron 193:113543. https://doi.org/10.1016/j.bios.2021.113543
10. Bosch I, Reddy A, de Puig H et al (2020) Serotype-specific detection of dengue viruses in a nonstructural protein 1-based enzyme-linked immunosorbent assay validated with a multinational cohort. PLoS Negl Trop Dis 14:e0008203
11. Bowen A, Casadevall A (2018) The role of the constant region in antibody-antigen interactions: redefining the modular model of immunoglobulin structure. Elsevier Inc.
12. Brazaca LC, dos Santos PL, de Oliveira PR et al (2021) Biosensing strategies for the electrochemical detection of viruses and viral diseases—a review. Anal Chim Acta 1159:338384. https://doi.org/10.1016/j.aca.2021.338384
13. Burrell CJ, Howard CR, Murphy FA (2017) Adaptive Immune Responses to Infection. Fenner White's Med Virol 65–76. https://doi.org/10.1016/b978-0-12-375156-0.00006-0
14. Cabral-Miranda G, Cardoso AR, Ferreira LCS et al (2018) Biosensor-based selective detection of Zika virus specific antibodies in infected individuals. Biosens Bioelectron 113:101–107. https://doi.org/10.1016/j.bios.2018.04.058
15. Cabral DGA, Lima ECS, Moura P, Dutra RF (2016) A label-free electrochemical immunosensor for hepatitis B based on hyaluronic acid–carbon nanotube hybrid film. Talanta 148:209–215. https://doi.org/10.1016/j.talanta.2015.10.083
16. Califf RM (2018) Biomarker definitions and their applications. Exp Biol Med 243:213–221. https://doi.org/10.1177/1535370217750088
17. Cassedy A, Parle-McDermott A, O'Kennedy R (2021) Virus detection: A review of the current and emerging molecular and immunological methods. Front Mol Biosci 8:1–21
18. Caygill RL, Blair GE, Millner PA (2010) A review on viral biosensors to detect human pathogens. Anal Chim Acta 681:8–15. https://doi.org/10.1016/j.aca.2010.09.038
19. Chen M, Qin R, Jiang M et al (2021a) Clinical applications of detecting IgG, IgM or IgA antibody for the diagnosis of COVID-19: A meta-analysis and systematic review. Int J Infect Dis 104:415–422. https://doi.org/10.1016/j.ijid.2021.01.016
20. Chen P-Y, Ko C-H, Wang CJ et al (2021b) The early detection of immunoglobulins via optical-based lateral flow immunoassay platform in COVID-19 pandemic. PLoS One 16:e0254486

21. Chen Z, Zhang Z, Zhai X et al (2020) Rapid and sensitive detection of anti-SARS-CoV-2 IgG, using lanthanide-doped nanoparticles-based lateral flow immunoassay. Anal Chem 92:7226–7231. https://doi.org/10.1021/acs.analchem.0c00784
22. Chien YW, Liu ZH, Tseng FC et al (2018) Prolonged persistence of IgM against dengue virus detected by commonly used commercial assays. BMC Infect Dis 18:1–7. https://doi.org/10.1186/s12879-018-3058-0
23. Chikkaveeraiah BV, Mani V, Patel V et al (2011) Microfluidic electrochemical immunoarray for ultrasensitive detection of two cancer biomarker proteins in serum. Biosens Bioelectron 26:4477–4483. https://doi.org/10.1016/j.bios.2011.05.005
24. Clark LC Jr, Lyons C (1962) Electrode systems for continuous monitoring in cardiovascular surgery. Ann N Y Acad Sci 102:29–45. https://doi.org/10.1111/j.1749-6632.1962.tb13623.x
25. Clem AS (2011) Fundamentals of vaccine immunology. J Glob Infect Dis 3:73–78. https://doi.org/10.4103/0974-777X.77299
26. Cubuk J, Alston JJ, Incicco JJ et al (2021) The SARS-CoV-2 nucleocapsid protein is dynamic, disordered, and phase separates with RNA. Nat Commun 12:1–17. https://doi.org/10.1038/s41467-021-21953-3
27. Dai L, Gao GF (2021) Viral targets for vaccines against COVID-19. Nat Rev Immunol 21:73–82. https://doi.org/10.1038/s41577-020-00480-0
28. Dawn Thompson J (2016) Introduction. In: Statistics for Bioinformatics. Elsevier, pp 3–25
29. De Decker S, Vray M, Sistek V et al (2015) Evaluation of the diagnostic accuracy of a new dengue IgA capture assay (platelia dengue IgA capture, bio-rad) for dengue infection detection. PLoS Negl Trop Dis 9:1–12. https://doi.org/10.1371/journal.pntd.0003596
30. de Eguilaz MR, Cumba LR, Forster RJ (2020) Electrochemical detection of viruses and antibodies: A mini review. Electrochem Commun 116:106762. https://doi.org/10.1016/j.elecom.2020.106762
31. de la Cruz Hernández SI, González Mateos S, Flores Aguilar H et al (2012) Evaluation of a novel commercial rapid test for dengue diagnosis based on specific IgA detection. Diagn Microbiol Infect Dis 72:150–155. https://doi.org/10.1016/j.diagmicrobio.2011.11.002
32. de la Escosura-Muñiz A, Maltez-da Costa M, Sánchez-Espinel C et al (2010) Gold nanoparticle-based electrochemical magnetoimmunosensor for rapid detection of anti-hepatitis B virus antibodies in human serum. Biosens Bioelectron 26:1710–1714. https://doi.org/10.1016/j.bios.2010.07.069
33. de Oliveira TR, Erbereli CR, Manzine PR et al (2020) Early diagnosis of Alzheimer's disease in blood using a disposable electrochemical microfluidic platform. ACS Sens 5:1010–1019. https://doi.org/10.1021/acssensors.9b02463
34. Deeks JJ, Dinnes J, Takwoingi Y, Davenport C, Spijker R, Taylor-Phillips S, Adriano A, Beese S, Dretzke J, di Ruffano LF, Harris IM, Price MJ, Dittrich S, Emperador D, Hooft L, Leeflang MM, den Bruel A (2020) Antibody tests for identification of current and past infection with SARS-CoV-2. Cochrane Database Syst Rev. https://doi.org/10.1002/14651858.CD013652
35. DeMers HL, He S, Pandit SG et al (2020) Development of an antigen detection assay for early point-of-care diagnosis of Zaire ebolavirus. PLoS Negl Trop Dis 14:e0008817. https://doi.org/10.1371/journal.pntd.0008817
36. Dip Gandarilla AM, Regiart M, Bertotti M et al (2020) One-step enzyme-free dual electrochemical immunosensor for histidine-rich protein 2 determination. RSC Adv 11:408–415. https://doi.org/10.1039/d0ra08729g
37. Dortet L, Ronat J-B, Vauloup-Fellous C et al (2021) Evaluating 10 commercially available SARS-CoV-2 rapid serological tests by use of the STARD (standards for reporting of diagnostic accuracy studies) method. J Clin Microbiol 59:e02342-e2420. https://doi.org/10.1128/JCM.02342-20
38. Douam F, Lavillette D, Cosset FL (2015) The mechanism of HCV entry into host cells. Prog Mol Biol Transl Sci 129:63–107. https://doi.org/10.1016/bs.pmbts.2014.10.003
39. Du L, He Y, Zhou Y et al (2009) The spike protein of SARS-CoV—A target for vaccine and therapeutic development. Nat Rev Microbiol 7:226–236. https://doi.org/10.1038/nrmicro2090

40. Faria AM, Mazon T (2019) Early diagnosis of Zika infection using a ZnO nanostructures-based rapid electrochemical biosensor. Talanta 203:153–160. https://doi.org/10.1016/j.talanta.2019.04.080
41. Flower B, Brown JC, Simmons B et al (2020) Clinical and laboratory evaluation of SARS-CoV-2 lateral flow assays for use in a national COVID-19 seroprevalence survey. Thorax 75:1082–1088. https://doi.org/10.1136/thoraxjnl-2020-215732
42. Fransen K, Vermoesen T, Beelaert G et al (2013) Using conventional HIV tests on oral fluid. J Virol Methods 194:46–51. https://doi.org/10.1016/j.jviromet.2013.08.004
43. Freiman JM, Tran TM, Schumacher SG et al (2016) Hepatitis C core antigen testing for diagnosis of hepatitis C virus infection: A systematic review and meta-analysis. Ann Intern Med 165:345–355. https://doi.org/10.7326/M16-0065
44. Furuyama W, Marzi A (2020) Development of an enzyme-linked immunosorbent assay to determine the expression dynamics of ebola virus soluble glycoprotein during infection. Microorganisms 8:1–10. https://doi.org/10.3390/microorganisms8101535
45. Gao J, Zhu X, Wu M et al (2021) IFI27 may predict and evaluate the severity of respiratory syncytial virus infection in preterm infants. Hereditas 158:1–14. https://doi.org/10.1186/s41065-020-00167-5
46. Grieshaber D, MacKenzie R, Vörös J, Reimhult E (2008) Electrochemical biosensors—Sensor principles and architectures. Sensors 8
47. Guillem P, Bustos RH, Garzon V et al (2021) A low-cost electrochemical biosensor platform for C-reactive protein detection. Sens Bio-Sensing Res 31:100402. https://doi.org/10.1016/j.sbsr.2021.100402
48. Han Y, Gao H, Xu J et al (2020) Innate and adaptive immune responses against microsporidia infection in mammals. Front Microbiol 11:1–11. https://doi.org/10.3389/fmicb.2020.01468
49. Horns F, Dekker CL, Quake SR (2020) Memory B cell activation, broad anti-influenza antibodies, and bystander activation revealed by single-cell transcriptomics. Cell Rep 30:905-913.e6. https://doi.org/10.1016/j.celrep.2019.12.063
50. Huang Y, Yang C, Xu X feng et al (2020) Structural and functional properties of SARS-CoV-2 spike protein: potential antivirus drug development for COVID-19. Acta Pharmacol Sin 41:1141–1149. https://doi.org/10.1038/s41401-020-0485-4
51. Hunsperger EA, Yoksan S, Buchy P et al (2014) Evaluation of commercially available diagnostic tests for the detection of dengue virus NS1 antigen and anti-dengue virus IgM antibody. PLoS Negl Trop Dis 8. https://doi.org/10.1371/journal.pntd.0003171
52. Huss R (2014) Biomarkers. In: Atala; A, Allickson J (eds) Translational regenerative medicine, 1st Editio. Elsevier Inc., pp 235–241
53. Idili A, Parolo C, Alvarez-Diduk R, Merkoçi A (2021) Rapid and efficient detection of the SARS-CoV-2 spike protein using an electrochemical aptamer-based sensor. ACS sensors 6:3093–3101. https://doi.org/10.1021/acssensors.1c01222
54. Ionescu RE, Cosnier S, Herrmann S, Marks RS (2007) Amperometric immunosensor for the detection of anti-west nile virus IgG. Anal Chem 79:8662–8668. https://doi.org/10.1021/ac0707129
55. Irshad M, Dhar I, Joshi YK (2006) Significance of hepatitis C virus core protein in the diagnosis of hepatitis C virus infection in different liver diseases. J Investig Med 54:478–483. https://doi.org/10.2310/6650.2006.05039
56. IUPAC (1992) Compendium of chemical terminology (the "Gold Book"), 2nd ed. Compiled by A. D. McNaught and A. Wilkinson. Blackwell Scientific Publications, Oxford
57. Jacob-Dolan C, Feldman J, McMahan K et al (2021) Coronavirus-specific antibody cross reactivity in rhesus macaques following SARS-CoV-2 vaccination and infection. J Virol 95:1–11. https://doi.org/10.1128/JVI.00117-21
58. Karadurmus L, Kaya SI, Ozkan SA (2021) Recent advances of enzyme biosensors for pesticide detection in foods. J Food Meas Charact. https://doi.org/10.1007/s11694-021-01032-3
59. Karunakaran C, Pandiaraj M, Santharaman P (2015) Immunosensors. Elsevier Inc.
60. Kaushik A, Yndart A, Kumar S et al (2018) A sensitive electrochemical immunosensor for label-free detection of Zika-virus protein. Sci Rep 8:3–7. https://doi.org/10.1038/s41598-018-28035-3

61. Kim SH (2017) ELISA for quantitative determination of hepatitis b virus surface antigen. Immune Netw 17:451–459. https://doi.org/10.4110/in.2017.17.6.451
62. Kroes ACM, Kox LFF (1999) Diagnostic techniques|detection of viral antigens, nucleic acids and specific antibodies. In: Encyclopedia of virology. Elsevier, pp 388–395
63. Kumar S, Bhushan P, Krishna V, Bhattacharya S (2018) Tapered lateral flow immunoassay based point-of-care diagnostic device for ultrasensitive colorimetric detection of dengue NS1. Biomicrofluidics 12:34104. https://doi.org/10.1063/1.5035113
64. Lang T, Lo C, Skinner N et al (2011) The hepatitis B e antigen (HBeAg) targets and suppresses activation of the Toll-like receptor signaling pathway. J Hepatol 55:762–769. https://doi.org/10.1016/j.jhep.2010.12.042
65. Letko M, Marzi A, Munster V (2020) Functional assessment of cell entry and receptor usage for SARS-CoV-2 and other lineage B betacoronaviruses. Nat Microbiol 5:562–569. https://doi.org/10.1038/s41564-020-0688-y
66. Li D, Li J (2021) Immunologic testing for SARS-CoV-2 infection from the antigen perspective. J Clin Microbiol 59. https://doi.org/10.1128/JCM.02160-20
67. Li HC, Lo SY (2015) Hepatitis C virus: virology, diagnosis and treatment. World J Hepatol 7:1377–1389. https://doi.org/10.4254/wjh.v7.i10.1377
68. Li Z, Yi Y, Luo X et al (2020) Development and clinical application of a rapid IgM-IgG combined antibody test for SARS-CoV-2 infection diagnosis. J Med Virol 0–1. https://doi.org/10.1002/jmv.25727
69. Lim RRX, Bonanni A (2020) The potential of electrochemistry for the detection of coronavirus-induced infections. TrAC—Trends Anal Chem 133:116081. https://doi.org/10.1016/j.trac.2020.116081
70. Liu F, Li L, Xu M Da et al (2020) Prognostic value of interleukin-6, C-reactive protein, and procalcitonin in patients with COVID-19. J Clin Virol 127:104370. https://doi.org/10.1016/j.jcv.2020.104370
71. Liu G, Rusling JF (2021) COVID-19 antibody tests and their limitations. ACS Sens 6:593–612. https://doi.org/10.1021/acssensors.0c02621
72. Liu KT, Liu YH, Lin CY et al (2016) Inflammatory molecules expression pattern for identifying pathogen species in febrile patient serum. Exp Ther Med 12:312–318. https://doi.org/10.3892/etm.2016.3323
73. Lopez AL, Adams C, Ylade M et al (2021) Determining dengue virus serostatus by indirect IgG ELISA compared with focus reduction neutralisation test in children in Cebu, Philippines: a prospective population-based study. Lancet Glob Heal 9:e44–e51. https://doi.org/10.1016/S2214-109X(20)30392-2
74. Low SL, Leo YS, Lai YL et al (2021) Evaluation of eight commercial Zika virus IgM and IgG serology assays for diagnostics and research. PLoS ONE 16:1–15. https://doi.org/10.1371/journal.pone.0244601
75. Ma Y, Yang J, Yang T et al (2020) Electrochemical detection of C-reactive protein using functionalized iridium nanoparticles/graphene oxide as a tag. RSC Adv 10:9723–9729. https://doi.org/10.1039/c9ra10386d
76. Mahmoudinobar F, Britton D, Montclare JK (2021) Protein-based lateral flow assays for COVID-19 detection. Protein Eng Des Sel 34:gzab010. https://doi.org/10.1093/protein/gzab010
77. Mahshid SS, Flynn SE, Mahshid S (2021) The potential application of electrochemical biosensors in the COVID-19 pandemic: A perspective on the rapid diagnostics of SARS-CoV-2. Biosens Bioelectron 176:112905. https://doi.org/10.1016/j.bios.2020.112905
78. Mattioli IA, Hassan A, Sanches NM et al (2021) Highly sensitive interfaces of graphene electrical-electrochemical vertical devices for on drop atto-molar DNA detection. Biosens Bioelectron 175:112851. https://doi.org/10.1016/j.bios.2020.112851
79. McBride R, van Zyl M, Fielding BC (2014) The coronavirus nucleocapsid is a multifunctional protein. Viruses 6:2991–3018. https://doi.org/10.3390/v6082991
80. Merkoçi A, Li C, Lechuga LM, Ozcan A (2021) COVID-19 biosensing technologies. Biosens Bioelectron 178:113046. https://doi.org/10.1016/j.bios.2021.113046

81. Meyer B, Torriani G, Yerly S et al (2020) Validation of a commercially available SARS-CoV-2 serological immunoassay. Clin Microbiol Infect 26:1386–1394. https://doi.org/10.1016/j.cmi.2020.06.024
82. Mistry DA, Wang JY, Moeser M-E et al (2021) A systematic review of the sensitivity and specificity of lateral flow devices in the detection of SARS-CoV-2. BMC Infect Dis 21:828. https://doi.org/10.1186/s12879-021-06528-3
83. Modrow S, Falke D, Truyen U, Schätzl H (2013) Viruses: definition, structure, classification. Mol Virol 17–30. https://doi.org/10.1007/978-3-642-20718-1_2
84. Mokhtarzadeh A, Eivazzadeh-Keihan R, Pashazadeh P et al (2017) Nanomaterial-based biosensors for detection of pathogenic virus. TrAC Trends Anal Chem 97:445–457. https://doi.org/10.1016/j.trac.2017.10.005
85. Moshe M, Daunt A, Flower B et al (2021) SARS-CoV-2 lateral flow assays for possible use in national covid-19 seroprevalence surveys (React 2): diagnostic accuracy study. BMJ 372:n423–n423. https://doi.org/10.1136/bmj.n423
86. Muller DA, Young PR (2013) The flavivirus NS1 protein: Molecular and structural biology, immunology, role inpathogenesis and application asadiagnostic biomarker. Antiviral Res 98:192–208. https://doi.org/10.1016/j.antiviral.2013.03.008
87. Nakatsuma A, Kaneda M, Kodama H et al (2015) Detection of HIV-1 p24 at attomole level by ultrasensitive ELISA with Thio-NAD cycling. PLoS One 10:e0131319
88. Naresh V, Lee N (2021) A review on biosensors and recent development of nanostructured materials-enabled biosensors. Sensors (Switzerland) 21:1–35. https://doi.org/10.3390/s21041109
89. Nawaz MH, Hayat A, Catanante G et al (2018) Development of a portable and disposable NS1 based electrochemical immunosensor for early diagnosis of dengue virus. Anal Chim Acta 1026:1–7. https://doi.org/10.1016/j.aca.2018.04.032
90. Nicholson LB (2016) The immune system. Essays Biochem 60:275–301. https://doi.org/10.1042/EBC20160017
91. Nuccetelli M, Pieri M, Gisone F, Bernardini S (2020) Combined anti-SARS-CoV-2 IgA, IgG, and IgM detection as a better strategy to prevent second infection spreading waves. Immunol Invest 1–13. https://doi.org/10.1080/08820139.2020.1823407
92. Ooi JSG, Lok SM (2021) How NS1 antibodies prevent severe flavivirus disease. Trends Biochem Sci 46:519–521. https://doi.org/10.1016/j.tibs.2021.03.005
93. Parate K, Rangnekar SV, Jing D et al (2020) Aerosol-jet-printed graphene immunosensor for label-free cytokine monitoring in serum. ACS Appl Mater Interfaces 12:8592–8603. https://doi.org/10.1021/acsami.9b22183
94. Pinyou P, Noguer T, Blay VBT-RM in BS (2021) Enzymes, aptamers, and their use in sensors. Elsevier
95. Pollock V (2007) Proteins. In: xPharm: the comprehensive pharmacology reference. Elsevier, pp 1–11
96. Prat CM, Flusin O, Panella A et al (2014) Evaluation of commercially available serologic diagnostic tests for chikungunya virus. Emerg Infect Dis 20:2129–2132. https://doi.org/10.3201/eid2012.141269
97. Prince HE, Matud JL (2011) Estimation of dengue virus IgM persistence using regression analysis. Clin Vaccine Immunol 18:2183–2185. https://doi.org/10.1128/CVI.05425-11
98. Qi X, Liu P, Liang P et al (2019) Dual-signal-biosensor based on luminescent bacteria biofilm for real-time online alert of Cu(II) shock. Biosens Bioelectron 142:111500. https://doi.org/10.1016/j.bios.2019.111500
99. Qureshi A, Niazi JH (2020) Biosensors for detecting viral and bacterial infections using host biomarkers: a review. Analyst 145:7825–7848. https://doi.org/10.1039/d0an00896f
100. Ramamurthy M, Kannangai R, Abraham AM, Sridharan G (2012) Viral infections in immunocompromised hosts. Proc Natl Acad Sci India Sect B Biol Sci 82:95–109. https://doi.org/10.1007/s40011-011-0008-7
101. Rashed MZ, Kopechek JA, Priddy MC et al (2021) Rapid detection of SARS-CoV-2 antibodies using electrochemical impedance-based detector. Biosens Bioelectron 171:112709. https://doi.org/10.1016/j.bios.2020.112709

102. Redegeld F, Hermann K, Ollert M, Ring J (2019) Antibody Detection. Nijkamp Parnham's Princ Immunopharmacol 233–241. https://doi.org/10.1007/978-3-030-10811-3_14
103. Rhouati A, Teniou A, Badea M, Marty JL (2021) Analysis of recent bio-/nanotechnologies for coronavirus diagnosis and therapy. Sensors (Basel) 21:1485. https://doi.org/10.3390/s21041485
104. Riedel S (2019) Predicting bacterial versus viral infection, or none of the above: Current and future prospects of biomarkers. Clin Lab Med 39:453–472. https://doi.org/10.1016/j.cll.2019.05.011
105. Santos A, Bueno PR, Davis JJ (2018) A dual marker label free electrochemical assay for flavivirus dengue diagnosis. Biosens Bioelectron 100:519–525. https://doi.org/10.1016/j.bios.2017.09.014
106. Santos C da C, Santos PCM, Rocha KLS et al (2020) A new tool for dengue virus diagnosis: optimization and detection of anti-NS1 antibodies in serum samples by impedimetric transducers. Microchem J 154:104544. https://doi.org/10.1016/j.microc.2019.104544
107. Schroeder HW Jr, Cavacini L (2010) Structure and function of immunoglobulins. J Allergy Clin Immunol 125:S41–S52. https://doi.org/10.1016/j.jaci.2009.09.046
108. Sena-Torralba A, Ngo DB, Parolo C et al (2020) Lateral flow assay modified with time-delay wax barriers as a sensitivity and signal enhancement strategy. Biosens Bioelectron 168:112559. https://doi.org/10.1016/j.bios.2020.112559
109. Shin M, Yoon J, Yi C et al (2019) Flexible HIV-1 biosensor based on the Au/MoS(2) nanoparticles/Au nanolayer on the PET substrate. Nanomater (Basel, Switzerland) 9:1076. https://doi.org/10.3390/nano9081076
110. Siew QY, Pang EL, Loh H-S, Tan MTT (2021) Highly sensitive and specific graphene/TiO_2 impedimetric immunosensor based on plant-derived tetravalent envelope glycoprotein domain III (EDIII) probe antigen for dengue diagnosis. Biosens Bioelectron 176:112895. https://doi.org/10.1016/j.bios.2020.112895
111. Staples JE, Monath TP, Gershman MD, Barrett ADT (2018) Yellow fever vaccines. In: Plotkin's vaccines, Seventh Ed. Elsevier, pp 1181–1265.e20
112. Stavra E, Petrou PS, Koukouvinos G et al (2018) Simultaneous determination of paraquat and atrazine in water samples with a white light reflectance spectroscopy biosensor. J Hazard Mater 359:67–75. https://doi.org/10.1016/j.jhazmat.2018.07.029
113. Sterlin D, Mathian A, Miyara M et al (2021) IgA dominates the early neutralizing antibody response to SARS-CoV-2. Sci Transl Med 13. https://doi.org/10.1126/scitranslmed.abd2223
114. Stone M, Bakkour S, Lanteri MC et al (2020) Zika virus RNA and IgM persistence in blood compartments and body fluids: a prospective observational study. Lancet Infect Dis 20:1446–1456. https://doi.org/10.1016/S1473-3099(19)30708-X
115. Strimbu K, Tavel JA (2010) What are biomarkers? Curr Opin HIV AIDS 5:463–466. https://doi.org/10.1097/COH.0b013e32833ed177
116. Svalova TS, Malysheva NN, Bubekova AK et al (2020) Effect of the method for immobilizing receptor layer on the analytical characteristics of a label-free electrochemical immunosensor for the determination of measles antibodies. J Anal Chem 75:254–261. https://doi.org/10.1134/S106193482002015X
117. Tang BM, Shojaei M, Parnell GP et al (2017) A novel immune biomarker IFI27 discriminates between influenza and bacteria in patients with suspected respiratory infection. Eur Respir J 49. https://doi.org/10.1183/13993003.02098-2016
118. Tarkkinen P, Palenius T, Lövgren T (2002) Ultrarapid, ultrasensitive one-step kinetic immunoassay for C-reactive protein (CRP) in whole blood samples: Measurement of the entire CRP concentration range with a single sample dilution. Clin Chem 48:269–277. https://doi.org/10.1093/clinchem/48.2.269
119. Taylor MW (2014) What is a virus? Viruses Man A Hist Interact 23–40. https://doi.org/10.1007/978-3-319-07758-1_2
120. Thangamuthu M, Santschi C, Martin OJF (2018) Label-free electrochemical immunoassay for C-reactive protein. Biosensors 8. https://doi.org/10.3390/bios8020034

121. Trombetta BA, Kandigian SE, Kitchen RR et al (2021) Evaluation of serological lateral flow assays for severe acute respiratory syndrome coronavirus-2. BMC Infect Dis 21:580. https://doi.org/10.1186/s12879-021-06257-7
122. Tso FY, Lidenge SJ, Poppe LK et al (2021) Presence of antibody-dependent cellular cytotoxicity (ADCC) against SARS-CoV-2 in COVID-19 plasma. PLoS ONE 16:1–12. https://doi.org/10.1371/journal.pone.0247640
123. Tucci M, Grattieri M, Schievano A et al (2019) Microbial amperometric biosensor for online herbicide detection: Photocurrent inhibition of Anabaena variabilis. Electrochim Acta 302:102–108. https://doi.org/10.1016/j.electacta.2019.02.007
124. Tyas AA, Raeni SF, Sakti SP, Sabarudin A (2021) Recent advances of hepatitis B detection towards paper-based analytical devices. Sci World J 2021. https://doi.org/10.1155/2021/6643573
125. Uliana CV, Riccardi CS, Yamanaka H (2014) Diagnostic tests for hepatitis C: recent trends in electrochemical immunosensor and genosensor analysis. World J Gastroenterol 20:15476–15491. https://doi.org/10.3748/wjg.v20.i42.15476
126. Valipour A, Roushani M (2017) TiO_2 nanoparticles doped with celestine blue as a label in a sandwich immunoassay for the hepatitis C virus core antigen using a screen printed electrode. Microchim Acta 184:2015–2022. https://doi.org/10.1007/s00604-017-2190-7
127. Valipour A, Roushani M (2018) Using boehmite nanoparticles as an undercoat, and riboflavin as a redox probe for immunosensor designing: ultrasensitive detection of hepatitis C virus core antigen. Anal Bioanal Chem Res 5:353–361. https://doi.org/10.22036/abcr.2018.122463.1195
128. Vasileva D, Badawi A (2019) C-reactive protein as a biomarker of severe H1N1 influenza. Inflamm Res 68:39–46. https://doi.org/10.1007/s00011-018-1188-x
129. Vasudevan D, Sreekumari S, Vaidyanathan K (2016) Proteins: structure and function. In: Textbook of biochemistry for medical students. Jaypee Brothers Medical Publishers (P) Ltd., pp 36–36
130. Vidarsson G, Dekkers G, Rispens T (2014) IgG subclasses and allotypes: From structure to effector functions. Front Immunol 5:520. https://doi.org/10.3389/fimmu.2014.00520
131. Voge NV, Sánchez-Vargas I, Blair CD et al (2013) Detection of dengue virus NS1 antigen in infected aedes aegypti using a commercially available kit. Am J Trop Med Hyg 88:260–266. https://doi.org/10.4269/ajtmh.2012.12-0477
132. Vollmer T, Diekmann J, Eberhardt M et al (2016) Monitoring of anti-hepatitis E virus antibody seroconversion in asymptomatically infected blood donors: systematic comparison of nine commercial anti-HEV IgM and IgG assays. Viruses 8:1–15. https://doi.org/10.3390/v8080232
133. Von Holle TA, Anthony Moody M (2019) Influenza and antibody-dependent cellular cytotoxicity. Front Immunol 10:1–8. https://doi.org/10.3389/fimmu.2019.01457
134. Warnecke JM, Lattwein E, Saschenbrecker S et al (2019) Added value of IgA antibodies against Zika virus non-structural protein 1 in the diagnosis of acute Zika virus infections. J Virol Methods 267:8–15. https://doi.org/10.1016/j.jviromet.2019.02.005
135. Wasik D, Mulchandani A, Yates MV (2018) Point-of-use nanobiosensor for detection of dengue virus NS1 antigen in adult aedes aegypti: A potential tool for improved dengue surveillance. Anal Chem 90:679–684. https://doi.org/10.1021/acs.analchem.7b03407
136. Whitford D (2005) Proteins: structure and function. Wiley, London, UK
137. Woof JM, Russell MW (2011) Structure and function relationships in IgA. Mucosal Immunol 4:590–597. https://doi.org/10.1038/mi.2011.39
138. Wubben R, Efstathiou C, Stevenson NJ (2021) Chapter one—the interplay between the immune system and viruses. In: Litwack GBT-V and H (ed) Hormones, regulators and viruses. Academic Press, pp 1–15
139. Yu HQ, Sun BQ, Fang ZF et al (2020) Distinct features of SARS-CoV-2-specific IgA response in COVID-19 patients. Eur Respir J 56:7–10. https://doi.org/10.1183/13993003.01526-2020
140. Yusa T, Tateda K, Ohara A, Miyazaki S (2017) New possible biomarkers for diagnosis of infections and diagnostic distinction between bacterial and viral infections in children. J Infect Chemother 23:96–100. https://doi.org/10.1016/j.jiac.2016.11.002

141. Zhan L, Granade T, Liu Y et al (2020) Development and optimization of thermal contrast amplification lateral flow immunoassays for ultrasensitive HIV p24 protein detection. Microsyst Nanoeng 6. https://doi.org/10.1038/s41378-020-0168-9
142. Zilla ML, Keetch C, Mitchell G et al (2021) SARS-CoV-2 serologic immune response in exogenously immunosuppressed patients. J Appl Lab Med 6:486–490. https://doi.org/10.1093/jalm/jfaa232

Miniaturized Electrochemical (Bio)sensing Devices Going Wearable

Lucas C. Faustino, João P. C. Cunha, Ana P. S. Andrade, Eliemy F. S. Bezerra, Roberto A. S. Luz, and Everson T. S. Gerôncio

Abstract Certainly, the idea of "wearing" a clock was unimaginable for the Egyptians who used their obelisks and sundials to measure time. However, thanks to technological advancement in human history, the advent of the wristwatch as the first wearable device was an inevitable one that brought a multitude of possibilities to humanity and to science. Currently, it is even common to go out on the streets and see people with smartwatches that, in addition to measuring the time, are able to check the individual's heart rate, blood pressure, and other health parameters. The new generation of wearable devices is on the way, focused on health assessment, security, forensic, environmental, and food quality control, with more emphasis on the first one. In this scenario, electrochemistry is playing an important role as the most promising detection system for this type of device, due to its easy miniaturization, outstanding (bio)sensor properties, simplicity, low power consumption, and low cost. With the increasing evolution of microfabrication and point-of-care technologies, wearable electrochemical sensors have been gaining great visibility due to their appeal and ability to analyze and monitor the user's health in real time, without the need to travel to a clinic and wait for days for the results. Therefore, this chapter will discuss an overview of wearable electrochemical sensors, highlighting components, fabrication, sensing techniques and the main aspects and possibilities to turn POC electrochemical (bio)sensing systems into wearable devices.

Keywords Wearable · Electrochemical devices · Sensors · Miniaturization

1 Introduction

Life expectancy has increased globally by more than 6 years in both men and women in the last two decades according to World Health Organization—WHO (from the

L. C. Faustino · A. P. S. Andrade · R. A. S. Luz · E. T. S. Gerôncio (✉)
Department of Chemistry, Federal University of Piauí, Teresina-PI CEP 64049-550, Brazil
e-mail: everson.thiago@ufpi.edu.br

J. P. C. Cunha · E. F. S. Bezerra · R. A. S. Luz · E. T. S. Gerôncio
Department of Chemistry, State University of Piauí, Teresina-PI CEP 64002-150, Brazil

F. N. Crespilho (ed.), *Advances in Bioelectrochemistry Volume 3*,
https://doi.org/10.1007/978-3-030-97921-8_3

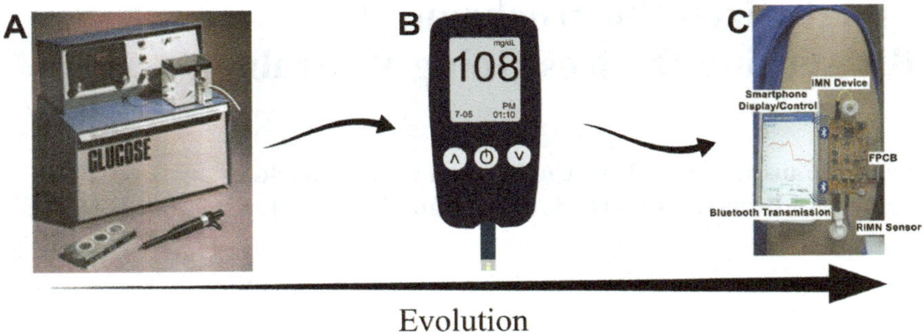

Fig. 1 Evolution of glucose electrochemical devices from **A** the first commercially successful glucose biosensor—YSI model 23A instrument (1975)—to the **B** more recent generation of glucometer with a pocketsize display, and **C** the future of such glucose monitoring devices with wearable continuous monitoring, wireless communication, and automatic insulin administration. Reproduced with permission from [20]

article "WHO methods and data sources for global burden of disease estimates 2000–2019" available in WHO website[1]). Part of the increasing life expectation is thanks to advancements in diagnostics and treatment of diseases research areas (among other things like lifestyle, sanitation, housing, education, etc.), which includes development of healthcare devices and biosensing systems [1]. The increasing personal desire of health awareness has led to a big explosion of wearable (bio)sensor devices development in the literature, supported by recent advances in miniaturized electronics [1, 2]. In this scenario, electrochemistry plays an important role and has gained tremendous attention over the past years for the fabrication of such miniaturized wearable (bio)sensing devices.

1.1 *Electrochemical Biosensors*

Biosensor is a system capable of recognizing an analyte through biochemical interaction and converting this event in a measurable signal that corresponds to the analyte concentration [3]. In other words, biosensors can be used to detect and quantify metabolites, electrolytes, biomarkers, drugs, and other analytes for diagnostics and/or continuous monitoring of an individual health (among other applications). Therefore, biosensors are becoming one of the most studied topics in science, leading to the next generation of medical care/self-care systems, especially due to its rapid, highly sensitive, and highly selective biochemical interactions, besides its relative low cost and easy miniaturization [4, 5].

[1] Global Health Estimates: Life expectancy and leading causes of death and disability. Retrieved from https://www.who.int/data/gho/data/themes/mortality-and-global-health-estimates.

In general, a biosensor is composed by four main parts (Fig. 1): (i) a biological recognition element (e.g., enzyme, antibody, and nucleic acids), which will selectively and specifically interact with the analyte; (ii) a transducer, responsible for translating the biochemical response into a measurable analytical signal proportional to the analyte concentration; (iii) a signal processor, in charge of measuring and amplifying the generated signal; (iv) finally, a data processing unit where the signal is "filtered" and converted into a more convenient measuring value like analyte concentration, presence or absence of a substance, etc. In some cases, parts (iii) and (iv) are combined into one resulting in a complete readout system [6].

Each part of the biosensor must be chosen accordingly to the desired application (which will be discussed later in this chapter), including the detection method. The past decades have shown increasing number of detection methods in biosensing systems based on a variety of transduction modes. Nowadays, electrochemical, optical, mass-sensitive, and thermal techniques are the most common transducers coupled to a biosensor [3]. Among then, electrochemical transducers have gained tremendous attention for biosensing applications in the last years, representing most publications in this field [7]. The increasing success of this type of biosensor is thanks to the exceptional attributes of electrochemical systems, such as high sensitivity, low cost, simplicity of instrumentation, ease operation, and possibility of miniaturization. Electrochemical biosensors use conductive electrodes as transducers modified with biorecognition elements like enzymes, antibodies, proteins, among others, to improve their selectivity. In most cases, the electrode is also functionalized with other (nano)materials (e.g., metal nanoparticles, carbon nanotubes, graphene, and its derivatives) to maximize its signal transduction and, sometimes, catalyze the intended reactions [8]. Some examples and principles of electrochemical biosensors will be given later in this chapter.

In general, the analytical signal obtained in electrochemical (bio)sensors is divided into four main categories: (i) chronoamperometric; (ii) voltammetric; (iii) potentiometric; and (iv) impedimetric analysis [9]. The first two detection principles involve generation of current through redox reactions that take place on the electrode surface, but with the former being under constant potential and the second being during a potential scanning [10]. These amperometric approaches are used when the analyte is an electroactive molecule or through some indirect mechanism using other electroactive species that participate on the electrochemical mechanism proportionally to the analyte's concentration. Potentiometry, unlike the amperometric analysis, measures potential difference between the reference electrode and a membrane on the working electrode due to some ion activity at the electrode interface. In other words, potentiometric systems determine the accumulated charge potential at an ion-selective working electrode when in presence of some specific ionic species, with no current passing through the system (zero current cell mode) [11]. Finally, electrochemical impedance spectroscopy (EIS) is a powerful electrochemical technique for measuring charge transfer resistance and/or capacitance on electrode–solution interfaces in response to changes of surface properties [12]. These impedance changes can be resulted from the binding of targets to bioreceptors (e.g., antibodies, nucleic acids) attached onto the working electrode surface in biosensing systems, for instance. More insights into electrochemical biosensors, strategies, and approaches can be found in some great review articles in the literature [2, 3, 7, 13].

1.2 The Role of Miniaturization

The importance of miniaturization in the sensing field comes from the concept that conventional analysis performed in a typical laboratory are normally laborious, time consuming, expensive, and require trained staff to carry out experiments [8, 14]. These peculiarities of typical laboratory analysis might turn it unavailable in some cases, especially in remote and/or developing regions. In this scenario, portable devices for on-site analyses may play an important role on the development of such technology, besides reducing cost of analysis and saving time in healthcare diagnostics field (among other applications) [2].

When comes to on-site diagnostics, portable systems should be not just miniaturized, but robust and autonomous, and consisting in integrated sample pretreatment, readout system, and data processing unit. This type of portable system is currently known as point-of-care (POC) or point-of-need (PON) testing devices [15]. POC/PON testing devices can be an auxiliary tool for telemedicine and continuous monitoring of patients, avoiding overcrowding in hospitals and excessive visits to the doctor. These characteristics make POC systems increasingly important, especially in this current pandemic scenario caused by the COVID-19.

The interest in POC technologies is not something new. In 1962, Clark et al. (1962) proposed an electrochemical device for continuous monitoring of glucose (glucometer), which ended up being one of the most successful POC electrochemical biosensor so far [16]. A glucometer consists of an amperometric device that fits in the palm of the hand, to which a glucose test strip (screen-printed electrodes modified with a specific enzyme for glucose) is attached. To complete the POC device, the glucometer is coupled to a pocket-size amperometric readout system containing all needed electronics for the electrochemical system to works. This type of POC device has been used by both clinical professionals and patients, where the last can perform their glucose monitoring in the comfort of their homes or at any other place. The commercial and scientific phenomenon of the glucometer as a POC device has exponentially expanded the interest in the development of new POC electrochemical biosensors for new purposes, such as cancer detection, for instance [17–19].

With the advent of nanotechnology and increasing evolution of micro- and nanoscale fabrication methods, complex instrumentation like conventional potentiostat systems is already being miniaturized and marketed for POC electrochemical analysis. Nowadays, even a smartphone or a wristwatch (smartwatch) can be used as a complete readout and data processing unit in POC electrochemical systems, which makes this type of analysis more and more affordable [21, 22]. Moreover, if we can wear a wristwatch as a POC fitness monitoring gadget, why not wearing it as our own personal diagnostic/monitoring healthcare device all the time. This is the future happening in the present, and science fiction becoming reality. Figure 1 shows the evolution of glucose electrochemical (bio)sensor toward miniaturization and beyond. In this chapter, we will focus on this type of electrochemical system presented in Fig. 1C and its challenges to become wearable.

1.3 Wearable Electrochemical Devices

The possibility of monitoring personal health (self-care) using relatively affordable everyday devices like smartphone or wristwatch, in a non-invasive way have led the scientific community to search for new analytical tools capable of performing *in-loco* accurate analysis directly on-body [23]. The research in this field was firstly driven to overcome major drawbacks of commercially available POC glucometers, which are the painful finger pricking and lack of continuous monitoring during day-life activities [24]. Thus, a lot of effort has been made to develop POC sensing technologies that could fit these criteria. For this to happen, the device should preferably be attached to the body (wearable) for continuous monitoring and must use a less invasive biological fluid (e.g., sweat, saliva, tears, and interstitial fluid). These efforts have led to many ideas and prototypes until the first commercially available real-time non-invasive continuous glucose monitoring was approved by FDA in 1999, which was named GlucoWatch® (discontinued in 2007) [25]. GlucoWatch® was a POC electrochemical device designed as a wristwatch to monitor glucose level in the interstitial fluid—ISF (located at the lowermost skin layer of dermis), which is extracted non-invasively by reverse iontophoresis [26]. Since then, other POC wearable devices have been developed in the literature for several applications, mainly designed for people in continuous therapeutic treatment and/or chronic diseases, which would of course benefit by this type of device.

Wearable sensing devices are defined as analytical tools that can be worn by an individual to keep tracking of their health (or fitness) and deliver the result by wireless communication (e.g., Bluetooth, NFC, and RFID) [27]. This type of device is normally directly attached to the individual's body, or it can be integrated into wearable objects like wristwatch or eyeglasses to provide clinically relevant data for care. Taking advantage of the exceptional attributes of electrochemical detection, already mentioned earlier in this chapter, besides its relative low consumption of electrical power, this type of device can become a powerful tool for healthcare analysis, among other applications [28]. Therefore, coupling electrochemical detection systems into wearable (bio)sensing technologies is a trending topic in current research related to analytical chemistry and healthcare management scientific fields.

Since its "beginning" in 2008, most works in the literature regarding wearable electrochemical (bio)sensors relied on the use of rigid electronic materials [23, 29]. However, following the recent advances of electronics industry on fabrication of high-resolution conductive patterns in flexible substrates, the research on wearable electrochemical (bio)sensors has shifted to develop stretchable and flexible electrodes/electronics [30]. Thus, this type of healthcare device could deform to adapt to the body's curvature or to the wearable gadget/clothes format without losing its electronic properties. Therefore, this major advance in fabricating flexible miniaturized electrodes and electronic systems became the starting point for the revolution of the wearable (bio)sensing field, especially considering electrochemical transducers.

Currently, the web-of-science® database contains over a thousand documents related to Wearable Electrochemical Sensors, being the first reported in early 90s about a potentially wearable glucose sensor [31]. However, only in the last decade there has been a significant increase in the number of publications and in the interest on developing wearable electrochemical sensors, as shown in Fig. 2. This recent rise of wearable electrochemical sensors publications and interest was largely due to the technological development that led to the concept of Internet of Things (IoT). With the popularization of smartphones and other mobile devices came the possibility of developing reliable and affordable wearable sensors that is more compact and capable of connecting with other devices through wireless technologies, such as Bluetooth. Thus, the user has the possibility of having access to the results instantly in the palm of his hand, in addition to enabling the instantaneous sending of data to a healthcare professional he trusts.

Considering the great relevance of this subject, in this chapter we are going to explore some of the main aspects and possibilities to turn POC electrochemical (bio)sensing systems into wearable devices. Initially, the development of such wearable electrochemical platforms will be discussed, focusing on miniaturization of electrodes, fabrication strategies, types of substrates, and samples (biological fluids). Later, operational details like transduction, data transmission and energy will be addressed, besides the applications of wearables electrochemical biosensing devices. And finally, considerations, perspectives, and challenges regarding the future of wearable electrochemical biosensing devices will be presented.

Fig. 2 Progress of publications related to wearable electrochemical biosensors in the last decade (2011–2021), using the Web of Science® search engine. Keywords used: "wearable" AND "electrochem*" AND "biosens*"

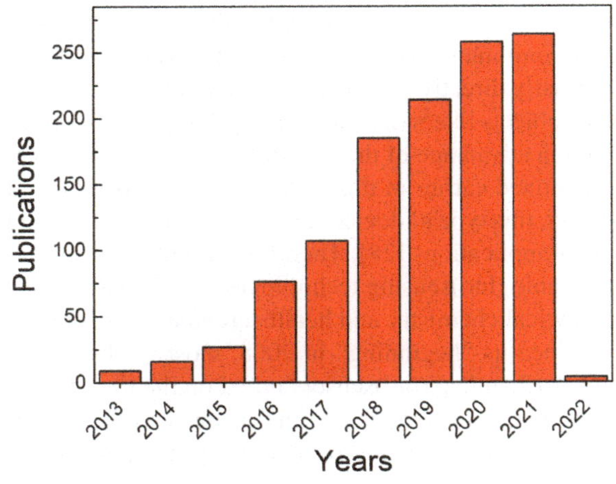

It is important to state that there are several review articles that have been available in the recent literature about this subject. Therefore, this chapter is not to include all great works recently published nor reach any specific conclusion on how to make wearable electrochemical biosensors, but to focus on basic structure, stimulate discussions, progresses, and present some challenges to overcome. For this reason, we are going to use only some representative recent literature to highlight important aspects for the context of this chapter and many relevant works in this field will unfortunately be overlooked.

2 Components

Miniaturized electrochemical systems for POC analysis are not quite something new, and they have been under development and available for many years, especially for single-use or short-term use analysis [8, 32, 33]. However, wearable electrochemical devices rely not only in miniaturization, but in flexibility, stretchability, and stability for long time usage as well, considering continuous monitoring of an individual's health. Therefore, proper instrumentation, fabrication, and (bio)sensing approaches should be considered to manufacture this type of wearable device.

2.1 Instrumentation

When comes to POC analysis, electrochemical potentiostats are one of the simplest benchtop instruments for miniaturization, especially if some unnecessary functions are removed for a given application. In fact, there are already several companies working on development and commercialization of miniaturized potentiostats for on-site analysis. PalmSens® (a Netherland's company) is one of those companies that has created several miniaturized electrochemical products for POC analysis, including one of the smallest potentiostats available in the market so far. These miniaturized electrochemical instrumentations are quite reliable, achieving similar results to those obtained using benchtop instruments. Therefore, thanks to the increasing evolution of electronic industry, miniaturized electrochemical instrumentation is in advanced stage, although it can be further improved to reduce costs and power consumption, besides better adaptation to wearable devices, for instance.

In addition to the commercially available miniaturized potentiostats, some works in the literature have shown the possibility of developing open-source DIY ("do-it-yourself") potentiostats [34], including with wireless communication and smartphone integration [35, 36]. This may open several possibilities for the creation of potentiostats that can fit into a specific wearable device. Delaney and co-workers have shown in 2013 another possibility of using a cellphone for potentiostatic control through the audio jack, for "instrument-free" sensing [36]. Although this was not set up for wearable (bio)sensing, it brings some interesting possibilities to be considered, especially if these new smartwatches can be used instead of cell phones.

2.2 *Electrodes*

Electrochemical systems in general require at least two electrodes (working and reference electrodes), together with the sample solution, to constitute the electrochemical cell. Working electrode (WE) is the electrode where the (bio)sensing event takes place; and reference electrode (RE) is the one needed to measure potential difference of the system and is also where the opposite reaction happens. A third electrode is often added to the system to avoid destabilization of the RE potential due to some changes on the electrode surface during analysis. This third electrode is known as counter or auxiliary electrode (CE) and is where the opposite reaction takes place in this three-electrode configuration [37].

There are some fabrication strategies presented in the literature that can be used to create miniaturized (2 or 3 electrodes) electrochemical systems and other conductive patterns [38–40]. Some of them are well described by Ferreira e co-workers (2019) in their review article about wearables electrochemical sensors for forensic and clinical application [28]. In general, wearable (bio)sensors are not rigid, which implies that the electrodes must be stretchable and bendable to avoid motion noise and for better adaptation to the body/wearable gadget curvature. Some authors have already described that the shape of the conductive electrode should be designed as waves (e.g., serpentine, island-interconnected, mesh structure, fractal, origami, and kirigami structural configurations) to improve stretchability and deformability [41, 42]. In this topic, we are going to summarize some of the main fabrication strategies employed in miniaturized electrochemical systems manufacturing, highlighting their advantages and disadvantages for wearable electrochemical (bio)sensors.

Screen-printing

Screen-printing is by far the most used method to produce miniaturized electrochemical systems for POC applications [43]. This technique uses a mesh with electrodes template designed using lithographic methods to allow conductive inks to pass only through the permeable parts and deposit onto the substrate at the desired pattern (Fig. 3A). Briefly, a high viscosity conductive ink is placed on top of the mesh and a blade is moved across the screen to fill up the permeable parts of the mesh. Then the blade moves backwards making the screen touch the substrate and transfer the ink. After some printing steps (e.g., WE and CE printing, RE printing, and isolating layer printing), the inks must be dried and sintered, normally at relative high temperature [44]. Screen-printing has the advantage of been susceptible to large-scale production of miniaturized electrochemical cells with different geometries and approaches [45]. Moreover, this fabrication technique is cost effective, robust, produce electrochemical devices with suitable performance, its highly repeatable and compatible with many substrates both rigid and flexible. For this reason, screen-printing has

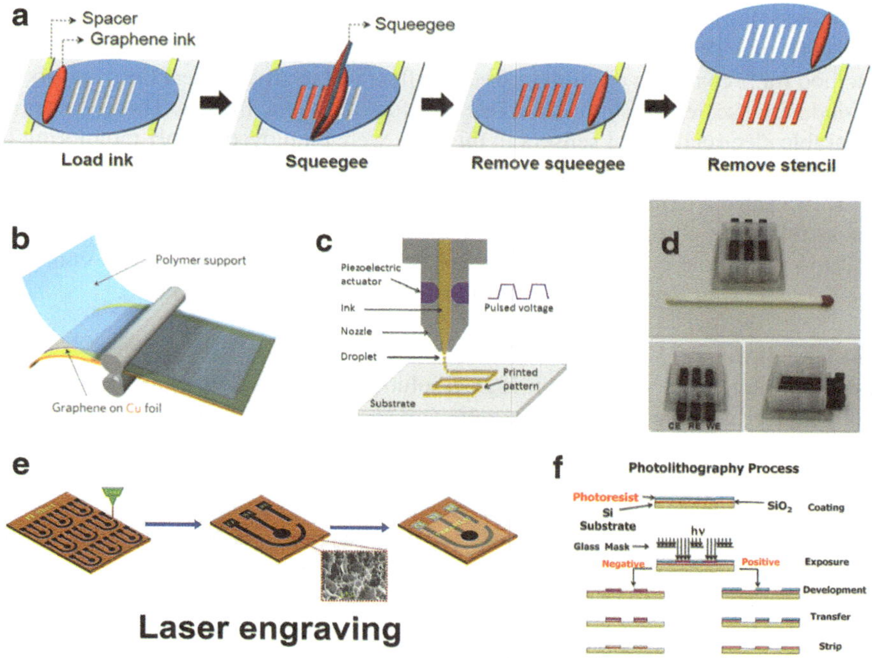

Fig. 3 Schematic representation of fabrication strategies for developing miniaturized electrodes. **A** Screen-printing of pristine graphene ink. Reproduced with permission from [63]. **B** Roll-to-roll (R2R) production of graphene films. Reproduced with permission from [64]. **C** Schematic of a conventional inkjet printing method. Reproduced with permission from [65]. **D** 3D printing of carbon-loaded PLA conductive filament. Reproduced with permission from [53]. **E** Laser-engraved graphene electrode production. Reproduced with permission from [66]. **F** Schematic of a photolithographic method. Reproduced with permission from [67]

revolutionized POC electrochemical (bio)sensing field, with hundreds of examples, including in wearable electrochemical devices manufacturing [44].

Metters et al. (2011) have stated in his review article that screen-printing would help establish the route for POC electrochemical sensing development from lab to market [46]. Even today, this technology is still on top as the most used fabrication method for this type of device, which has led to the development of some exiting wearable electrochemical (bio)sensors [44]. However, there are some limitations that should be considered when using screen-printing as fabrication technique, such as the laborious and expensive prototyping, template screen requirement for each printing layer, and unsuitability for curved surfaces (unless the surface is bendable and can be turned into its planar form).

Roll-to-roll gravure printing

Roll-to-roll (R2R) gravure printing is another large-scale production method that has been used to create wearable electrochemical (bio)sensors (Fig. 3B). Using this technique, Bariya et al. (2018) was able to prepare about 150,000 electrodes on a 150 m flexible PET roll [47]. This achievement highlights the possibility of using this method for preparation of miniaturized and flexible electrochemical (bio)sensors on an industrial scale, which makes it one step forward to the market. The as-prepared wearable electrochemical sensors in this case were employed for pH monitoring in sweat during exercise activities.

R2R is also a template-based method and, as screen-printing, it has some drawbacks related to prototyping, besides its unsuitability for rigid substrates due to its roll-to-roll routine procedure. This latest is not really a drawback when comes to wearable sensing devices, where flexibility is of paramount importance.

Inkjet printing

Inkjet printing is a well-known high-resolution non-contact technique that does not require template for the preparation of a conductive pattern, and therefore, it has a fast and simple prototyping, unlike the ones previously mentioned [48]. Moreover, inkjet printing offers relative low cost, scalability, and reduced waste of printing materials, which makes this technique highly eco-sustainable. Inkjet printing is suitable toward different types of low viscosity inks, such as bioactive fluids, catalytic materials, isolating layer, and conductive patterns. This technique works by depositing microdroplets directly onto the substrate with high accuracy, using different jetting mechanisms (e.g., piezoelectric, thermal, and electrohydrodynamic actuation), followed by drying and sintering processes (Fig. 3C). Some works in the literature have already shown the use of such technology for the preparation of wearable electrochemical/electronic (bio)sensing devices [49, 50].

Inkjet printing can be performed in both rigid and flexible substrates, although it cannot be printed in curved surfaces. Therefore, the well-suitability of this method toward flexible substrates and its high precision makes it a promising candidate for fabrication of wearable electrochemical (bio)sensors, especially because high-resolution and precision are required in this type of device. Nevertheless, there are some challenges in inkjet printing that hinders its possibility to becoming industrial, such as the slow speed of printing and the strict ink rheology to ensure good printing quality and to avoid nozzle clogging [51].

3D printing

3D printing is a recent template-free deposition method that is just at the beginning of its journey to becoming an important tool in analytical chemistry [52]. This method consists in stacking layers of the desired material in a controllable and very precise way to form a 3D structure, previously designed in specific software (Fig. 3D). Compared to the previously mentioned printing techniques, 3D printing is more compatible with different substrates, both rigid and flexible, inexpensive, versatile and allows fabrication of fully printed POC devices [53]. In other words, 3D printers can be used to print not just the physical parts of the wearable device, but also the electrodes, due to some new conductive materials that have been adapted to this technology.

Up to now, there are several 3D printing technologies in the market. Among them, the Fused Deposition Modeling (FDM) is the most widely employed for the fabrication of electrochemical systems, especially due to the low cost of printers and materials and its easy processability [52]. Despite its advantages and promising feature for electroanalytical chemistry, 3D printing has some drawbacks related to printing speed, scalability, and deposition of different materials (e.g., structural, and conductive ones) in the same device.

Most 3D printers that can work with more than one printing material at the same time are quite expensive, which reduces cost effectiveness of such technology. Nevertheless, this technology is in constant evolution and still quite promising for electroanalytical purposes, making its way to wearable electrochemical (bio)sensing platforms. As an example, Padash et al. [54] have recently shown the development of a 3D-printed wearable microfluidic platform for sweat analysis integrated with a commercially available screen-printed electrode. Although the device was not entirely 3D printed, the idea of using this technology to fabricate flexible on-body devices was planted. Katseli and co-workers [53] have demonstrated a fully 3D-printed miniaturized electrochemical cell, but the design of the device was not suitable for wearable electrochemical analysis. Later in 2021, the same group published an article of a fully 3D-printed wearable smartphone-assembled 3D-printed wearable electrochemical ring for non-enzymatic self-monitoring of glucose [55]. Therefore, it is believed that more 3D-printed wearable (bio)sensing devices will be witnessed in the literature in near future.

Laser engraving

Laser engraving is a technique discovered in 2014 to prepare 3D porous graphene-based conductive electrodes by direct laser writing using CO_2 laser onto a polymeric carbon-based substrate [56]. This technique was primarily used to reduce graphene oxide (GO) in specific parts of the GO film, to create the desires reduced GO (rGO) conductive pattern (Fig. 3E). Later, this technique was used to convert commercial polymers like polyimide into porous graphene by direct laser scribing with a commercially available CO_2 infrared laser system [57, 58]. Zhao and co-workers (2020) have recently demonstrated the use of laser engraving to create a flexible plant-wearable biosensor for in situ pesticide analysis [59]. Yang et al. (2020) went further and show the possibility of creating not just the highly sensitive electrochemical part of the device, but also the physical sensor and microfluidic platform [60]. In other words, the authors created all main parts of the wearable (bio)sensor device using laser engraving technique, highlighting the possibilities of using such technology. Moreover, they were able to demonstrate continuous monitoring of body temperature, respiration rate, uric acid, and tyrosine at low limit of detection.

As it can be seen, laser engraving has the advantage of rapid prototyping, geometry free, reagentless, and it can be performed in ambient air in both rigid and flexible substrates. However, this laser engraving technology is quite limited in terms of substrate and electrode materials that can be used, since only carbon-based conductive pattern is formed. Therefore, the use of complementary techniques is needed in this case to produce a complete miniaturized electrochemical cell in wearable devices, especially to print the RE and the isolating layer of the system.

Lithographic methods
Lithography is an old printing method invented in eighteenth century to create artwork materials based on immiscibility between oil and water. Nowadays, lithographic techniques are used in several areas, including electronics industry, and technology (Fig. 3F). Soft lithography, photolithography, and e-beam are common examples of this printing method, all presenting incredible precision and resolution, resulting in high-performance electrochemical systems with great reproducibility [38]. Lithographic methods can produce micro-scale electrochemical/electronic patterns that can be quite advantageous for POC and wearable (bio)sensing devices [61]. As an example, Ma et al. (2020) have demonstrated the use of photolithographic method to create flexible micropatterned wearable electrochemical glucose sensor [62]. Despite these advantages and possibilities, lithographic methods are very expensive, require trained staff and a clean-room facility, besides having laborious fabrication procedure. These inconvenient peculiarities of lithographic methods rise the costs of (bio)sensing devices production and ends up concentrating the research in this field only in well-resourced laboratories.

There are many other fabrication strategies that could be mentioned here for fabrication of miniaturized electrodes like stamp transfer, toner-printing [39], and pencil-based technologies [68–71]. However, this would increase the size of this chapter and probably run away from the proposed scope. Therefore, we decided to discuss only few techniques that holds great promising for wearable electrochemical (bio)sensing, which doesn't mean that these other methods are not suitable or promising for fabrication of this type of device. Stump transfer, for instance, has been demonstrated as an interesting approach for fabrication of flexible electronic/electrochemical devices, with possibility of creating conductive patterns in non-planar surfaces [72]. Windmiller et al. (2012) also published an article demonstrating the use of stamp transfer in wearables electrochemical (bio)sensing fabrications, including direct printing onto the skin [73].

2.3 Substrates

The material chosen as platform for manufacturing wearable electrochemical (bio)sensors is also crucial as component of the device. There are many materials, both rigid and flexible, that can be used for the purpose of being a support where electrodes are printed, microfluidic channels are formed and actuators/electronics are coupled, which obviously depends on the desired application [27, 45]. Generally, these platforms should present some particularities for the construction of wearable sensors, such as biocompatibility, mechanical resistance, cost-effectiveness, and flexibility in some specific cases [74, 75]. The literature of wearable devices is full of flexible substrates being used as platforms for the fabrication of such devices, such as polydimethylsiloxane (PDMS), poly(methyl methacrylate) (PMMA), polyimide (PI), polyethylene terephthalate (PET), among others [74]. But do these substrates really need to be flexible to be used in wearable systems?

Well, as mentioned earlier, GlucoWatch® was the first wearable electrochemical glucose monitor that received an approval from FDA, and it wasn't flexible. Designed as a wristwatch, GlucoWatch® used a miniaturized electrochemical system that could be worn in the lower arm and monitor glucose levels non-invasively [25, 26]. Although the electrodes of the amperometric system in this device were rigid, they were separated from the curved skin by flexible gel pads embedded with Gox enzyme, which makes it suitable for wearable biosensing directly onto the skin. Nevertheless, fabrication of wearable devices in flexible substrates is still a trending topic in this field, especially to avoid the use of external gadgets, and focus on devices that can be worn directly on-body for more comfort.

Under most conditions, using a wearable device leads to an inevitable noise provoked by motion and deformation that can bias the analytical signal. This noise can compromise the sensing performance of the electrochemical device, which was demonstrated by Foster and co-workers in [76]. Zhu et al. (2018) have also shown that wearable electrochemical sensors are very sensitive to low magnitude physical motion, which in his case was used to monitor vital signs like blood pressure, respiration, muscle activities and artificial tactile sensation (motion based physical parameters) [77]. However, for electrochemical detection of analytes in a biological fluid (or in the environment), being motion sensitive means that the sensor will respond not just to the analyte, but to the normal daily human physical activities as well, making it difficult to interpretate. For this reason, a lot of effort has been made to design wearable flexible electrochemical (bio)sensors that are not just bendable, but stretchable as well, aiming to maintain its electrochemical sensing performance [78, 79].

Several works in recent literature have shown that wearable sensing flexible devices for on-body analysis requires not just wave-shaped miniaturized electrodes (as mentioned earlier), but substrates that are extensible, providing high mechanical robustness. In this scenario, PDMS has been spotted out as a promising material to be used as platforms for the fabrication of wearable electrochemical devices, especially due to its flexibility, stretchability, hydrophobicity, dielectric properties, biocompatibility, and so on [74]. However, it is important to mention that the substrate must also be resistant to the process of electrodes fabrication to be used as platform of this type of device, which sometimes requires high temperature. In this case, PDMS suffers from its poor thermal stability, and PI rises as an interesting polymeric candidate because of its heat resistance (PI can withstand 300 °C of continuous heat) [80].

Finally, there are many other extensible substrates that have been used for manufacturing wearable systems, namely fibers, papers, and textiles (e.g., cotton, nylon, and wool). For more detail about this, some review articles on wearable/flexible electrochemical sensor are available in the literature, where the authors dedicated a topic or more for describing flexible substrates in this type of devices [81–84].

2.4 Samples (Biological Fluids)

Samples are not really part of the detection system, but the type of sample chosen for the wearable biosensor dictate the whole structure of the device to be manufactured, namely design, materials, sampling, and transduction. Samples in wearable sensor are normally biological fluids that are divided in two main categories: (i) non-invasive; and (ii) minimally invasive. Sweat, saliva, and tears are examples of non-invasive biological fluids, meanwhile interstitial fluid is considered by some authors as minimally invasive [24, 85, 86].

Sweat
Sweat has been the most used biological fluid to work as a model for development of wearable systems, especially due to its ease extraction and wealth of (bio)chemical information [87]. Some review articles have recently been published summarizing the advances on development of wearable devices for sweat analysis (also known as epidermal sensors) [82, 88]. Although sweat is a rich biological fluid, containing several important ions and metabolites, it has some drawbacks for wearable sensing, such as possibility of contamination and/or altered composition because of remaining species present in the epidermis, and the need of thermal heat or physical motion to be excreted from the body. Fortunately, there are other ways of sampling sweat like by chemical inducing [86], which will be addressed in the sampling topic.

Saliva
Saliva is another rich biological fluid that has the advantage of being easily extracted in sedentary condition and in large amounts, containing multiple biomarkers that can be used to access an individuals' health [24]. Kim et al. (2015) have demonstrated the development of a mouthguard-based wearable electrochemical biosensor, for real-time monitoring of some relevant metabolites like glucose and uric acid [89]. Nevertheless, there are some challenges that need to be overcome when using saliva as biological fluid sample. Constant variation of salivary contents due to daily meals, risk of electrodes biofouling and diluted analytes when compared to blood sample are some of those challenges.

Tears
Tears has been considered as another important biological fluid for non-invasive analysis. This biofluid is composed by various relevant biomarkers for healthcare assessment, such as glucose and uric acid, besides some electrolytes. One of the most recognizable examples of wearable devices for measuring specific analytes in tears is the Google contact lens [90]. This device was developed to measure glucose in tears using a miniaturized electrochemical system coupled to a wireless chip and an antenna. More articles have been published using the same concept of wearable electrochemical system in contact lenses for monitoring the glucose [91]. Despite the possibilities of using this type of biofluid for wearable sensing (not just for glucose monitoring, but for other metabolites and/or electrolytes as well), the risk of irreversibly damaging the eye is something that should be considered when developing

this type of wearable device. Moreover, some individuals are more susceptible of eye irritation than others when using contact lenses, which may interfere on the analytes concentration and hinder the analysis.

Interstitial fluid (ISF)

Interstitial fluid (ISF) is a biological fluid located at the lowermost skin layer of dermis, surrounded by cells and in constant equilibrium with the blood. For this reason, ISF is considered a quite attractive biofluid for wearable (bio)sensing since its composition is closely related to the blood [85]. The problem within this ISF biofluid is that it is not secreted spontaneously through the skin, but it can be sampled by reverse iontophoresis or by using microneedles in a non-invasive and painless way. Bandodkar et al. (2015, 2018) have shown a tattoo-based non-invasive glucose monitoring wearable system that uses reverse iontophoresis to access the ISF without damaging the skin [86, 92]. On the other hand, Madden and co-workers (2020) have recently published a review article related to biosensing in dermal ISF using microneedle based electrochemical devices [85]. Nevertheless, even for a minimally invasive system, there is a possibility of skin irritation or bruising appearance during usage, which might be considered uncomfortable and inconvenient for the user [93].

Although energy and communication systems are also components of a wearable electrochemical device, we decided to address those in the next section, together with data transmission, and sampling.

3 Operational Details

Heikenfeld et al. (2018) stated in his review article that probably many wearable chemical sensors will be chemical-to-electrical or electrochemical in nature in the future [94], especially due to their inherent miniaturization, low power requirement, simplicity, low-cost fabrication, among other already mentioned in this chapter. Miniaturization of reliable electrochemical systems are on the way, with significant progresses achieved in the last years, including for wearable applications [27, 32]. However, there are still some critical challenges that need to be addressed for continuous on-body monitoring. In this section, we are going to briefly discuss about some of those challenges, starting with the transduction part, then data transmission, followed by energy and finally sampling.

3.1 Electrochemical Methods/Transduction Under Wearable Settings

In general, analytical devices feature a recognition element, a signal transducer and a signal processing and data storage unit. The electrochemical transduction system has the function of converting a chemical or physical event provided by the interaction of the analyte with the recognition element into a measurable electrical signal [95, 96]. In electrochemical sensors, the working electrode functions as a transducer and measurements are carried out giving a signal of electrical nature, which can be potential, current, or impedance [94]. The type of electrical signal is important to develop a proper wearable electrochemical device for a given application, which clearly depends on the chosen sensing approach.

Potentiometric transduction

Potentiometry is the most used technique for wearable electrochemical (bio)sensing, mostly for monitoring electrolytes concentration (Na^+, Cl^-, Ca^{2+}, K^+, $NH_4{}^+$, Zn^+, Cd^{2+}, Pb^{2+}, Cu^{2+}, Hg^+) in sweat and other biological fluids [97]. Measuring potential difference between the ion-selective working electrode[2] (ISE) and the reference electrode is simpler and requires less apparatus/signal processing them other electrochemical approaches, besides ensuring continuous monitoring with fast analytical response. Moreover, ISE is less sensitive to (bio)fouling during measurement due to its charge-transduction mechanism, which highlights the advantages of potentiometry for on-body electrochemical (bio)sensing [94]. A review article on wearable potentiometric sensors was recently published by Parrilla and co-workers (2019) highlighting all advantages and possibilities within using such type of sensor [97]. However, it is important to state that potentiometric technique is limited to ionic analytes and therefore cannot be employed for monitoring uncharged metabolites, such as hormones, proteins, and some other biomarkers. For this reason, other electrochemical techniques should be considered depending on the type of application.

Amperometric transduction

Unlike potentiometric analysis, amperometry relies on faradaic events (redox reactions) that take place on the electrode surface. Therefore, analytes should be electroactive, otherwise an indirect mechanism must be presented in which some electroactive specie has its concentration on the electrode surface affected by the presence of the analyte. As already mentioned in this chapter, there are two ways of performing amperometric analysis: (i) one in which a fixed voltage is applied, and the current is monitored over time (chronoamperometry); and another (ii) where a time-dependent potential is applied to an electrochemical cell, and the resulting current is recorded as a function of that potential (voltammetry). In general, chronoamperometric analysis are simpler them the voltammetric ones, especially because its signal processing, allowing continuous on-body real-time monitoring of a given analyte [37].

Nevertheless, both amperometric methods are susceptible to fouling during analysis, which represents one of the most challenging aspects of electrochemical (bio)sensing in wearable systems [27]. Electrode fouling involves passivation of the electrode surface by a fouling agent, which might be in the sample, or is the product of a redox reaction, or even the analyte itself. This type of passivation generally compromises sensitivity, limit of detection, and reproducibility, i.e., the overall

[2] Coated-wire electrodes (CWEs) and ion-selective field effect transistors (ISFETs) are the other two types of working electrodes that can be employed in potentiometric analysis [37].

reliability of the sensor. Therefore, development of antifouling strategies for wearable electrochemical (bio)sensor devices are of the utmost importance considering continuous on-body monitoring. Some recent review articles have been published focused on strategies to mitigate electrode fouling, such as the usage of coating materials on the electrode surface or performing electrochemical activation [98, 99].

In amperometric analysis, working electrodes usually don't have selective membranes like in potentiometry, instead, (bio)recognition elements (e.g., enzymes, antibodies, and nucleic acids) are normally added to the electrode to ensure selectivity. Among the biorecognition elements, enzymes have been the most widely used in miniaturized/wearable electrochemical systems, with some remarkable success already in the market, as in the case with traditional glucometers [8]. However, the use of enzymes may result in poor stability under physiological conditions and variable enzyme activity due to changes in the environment when used for continuous monitoring and in wearable settings [75]. Therefore, in this case, the use of more stable recognition elements (with aptamers being one of the most promising for wearable electrochemical applications [8, 100]), or even a non-enzymatic approach should be considered for long-term on-body analysis [101].

Impedimetric transduction

Electrochemical impedance spectroscopy (EIS) is a powerful technique for the investigation of both bulk and interfacial electrical properties of the electrode systems. EIS is a non-destructive technique that has high sensitivity and doesn't require labeling, mediators, recognition elements, nor membranes, making it a promising alternative for wearable (bio)sensing applications [102]. However, there are some key challenges that hinders the use of EIS in continuous real-time monitoring, such as the linearity, stability, and causality requirements, for instance [103]. This means that the operating procedure is crucial for the success of impedance measurements. Moreover, EIS normally requires long analysis time, which also doesn't match with the wearable (bio)sensor modality. Nevertheless, Nah and co-workers (2021) have just demonstrated the development of a wearable impedimetric immunosensor for noninvasive sweat cortisol detection [104]. The impedimetric cortisol sensor showed great performance with low limit of detection (3.88 pM) and good selectivity, highlighting the possibilities of using such technique.

3.2 Data Transmission/Communication

Wearable sensors should preferably be able to communicate wirelessly with other electronic devices (e.g., smartphone, smartwatch, tablet, and notebook), so relevant data can be comfortably sent to a processing/monitoring unit without interfering with the persons daily activities [23, 27]. Nowadays, there are various wireless technologies and networks that allow people to send data to others without the aid of cables or wires. However, the monitoring and transferring of data are significantly dependent on the network in which the sensors are connected. Therefore, the type of wearable sensor may define which wireless technology should be employed in each case [23].

Selecting a particular wireless network for wearable electrochemical sensors depends on the cost, power consumption, number of sensors, and trans-reception range, among other factors [23]. Therefore, choosing these protocols in sensor development is an important task, as it affects system performance and characteristics. There are different wireless technologies that can be implemented in wearable platforms that are based on a wide network of connected things (e.g., sensors, actuators, identifiers, and communication units) [27]. In this scenario, Bluetooth, ZigBee, Radio-Frequency Identification (RFID), and Near-Field Communication (NFC) stand out as the most common wireless transmission technologies for developing wearable electrochemical (bio)sensors [27]. However, considering its low cost, less hardware requirement, high compatibility, and low energy consumption, Bluetooth is still the most used transmission protocol in this type of device [105].

Wireless protocols for wearable sensors can be classified into active and passive depending on the power supply needed for the devices to function. Bluetooth and ZigBee are examples of active systems, which require power supply, while the passive ones do not depend on energy (e.g., RFID and NFC) [27]. Figure 4 shows some

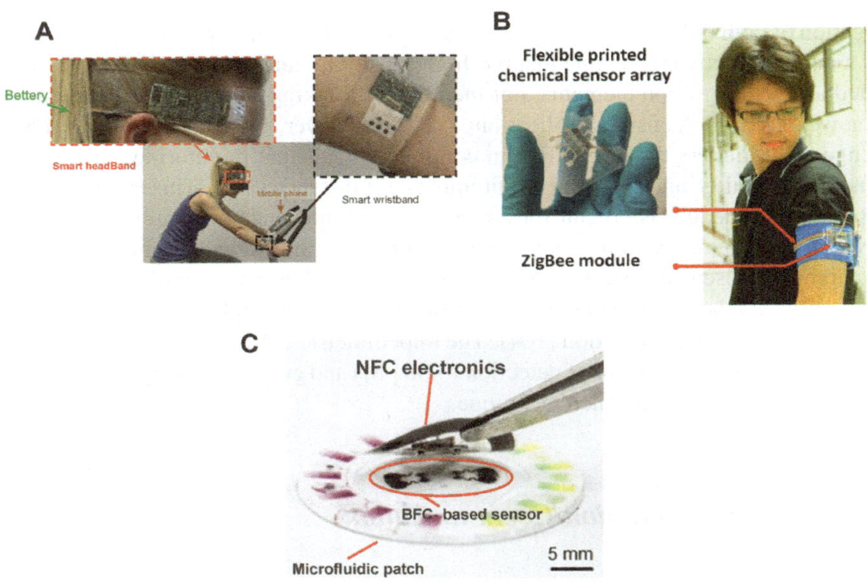

Fig. 4 ACTIVES: **A** Bluetooth-integrated wearable sensors with "smart headband" and "smart wristband" for sweating monitoring. Adapted and reproduced with permission from [106]. **B** Prototype of a wearable electronic nose based on the use of flexible inkjet-printed chemical sensor array integrated in the ZigBee wireless network. Adapted and reproduced with permission from [107]. PASSIVE: **C** Wearable optical sensor with NFC transmission components. Adapted and reproduced with permission from [95]

examples of active and passive wearable sensors that have been developed.

Sensors based on Bluetooth and ZigBee technologies have active transceivers (transmitter-receiver) capable of communicating with a reader device or other sensors in a network with 10–100 m of range [27]. Therefore, active transmitting wearable sensors need to have at least a battery or other power source to power both the sensor circuitry and data transmission. Despite the distance achieved by active technology sensors, such protocol can acquire interference from other wireless devices in the same range, which limit its applications. On the other hand, passive transmission systems for wearable electrochemical sensors, i.e., RFID and NFC protocols, communicate with reader devices through electromagnetic modulations generated by the reader. In this case, the reader must be only a few centimeters, <5 cm (NFC) up to several meters <15 m (RFID), from the sensor, which reduces interferences from outsider electronics.

Since there is no energy source in the passive protocol systems, the data cannot be registered autonomously and continuously. In fact, these devices generally require human intervention to bring the reader and sensor closer for data record, limiting the perspective of wearable sensing for this type of transmission [108]. Nevertheless, passive systems remain interestingly viable for wearable electrochemical sensing, especially due to their simple configurations, compact size, proximity sensors operational mode, and almost no power requirement. Table 1 shows the comparison of the as mentioned dominant network transmission protocols commonly used in wearable sensors.

There are other network transmission technologies also available for wearable electrochemical devices, such as Infrared Data Association (IrDA), Ultra-wideband (UWB), and Wi-Fi. IrDA is a wireless communication system (<10 cm) with high transmission data rate, but that need to be kept aligned with the sensor for a proper data transmission, which makes it unfeasible for wearable monitoring systems. The UWB offers a very high data rate at a very low power spectral density, which protects from possible interference with other radio waves. Although UWB technology holds promise for applications in medical monitoring and imaging [117, 118], the complexity and limited availability of commercial UWB systems make it impractical for wearable systems. Finally, Wi-Fi technology holds great promise for applications in wearable electrochemical sensor systems. However, its extremely high-power consumption and complex configuration make it inefficient for long-term wearable monitoring systems where battery life is indispensable [1].

3.3 Energy

Wearable electrochemical sensors are usually employed for sensing multiple simultaneous parameters, performing complex data analysis, and communicating with other sensors, devices, besides transmitting data, being therefore an energy dependent device [119]. As the demand for these devices grows, so does the requirement for relevant, compact, and efficient energy sources. Compact-size batteries are one

Table 1 Comparison and features of wireless protocols commonly used in wearable sensor applications

Protocol	Transceiver type	Radio band	Data rate	Range	Features	References
Bluetooth	Active radio, battery powered	2.4 GHz	1 Mbps (BLE) 24 Mbps (HS)	10–100 m	Very high data rate, good indoor range, Smartphone compatible, good for real time data transfer	[89, 109, 110]
ZigBee	Active radio, battery powered	868 MHz, 915 MHz, 2.4 GHz	<250 kbps	10–100 m	Very low power, good indoor range, versatile network topologies, good for local sensor/actuator networks and short bursts of data	[107, 111]
RFID	Passive transponder, batteryless or battery assisted	128–135 kHz, 13.56 MHz, 860–960 MHz	<100 kbp	<15 m, frequency dependent	Ultra-low-power, short range or contactless, very low data rate, smartphone compatible (13.56 MHz only), good for e-labels and small file transfers	[112–114]
NFC	Passive transponder, batteryless or battery assisted	13.56 MHz	<424 kbps	<5 cm	Ultra-low-power, contactless only, Smartphone compatible, good for secure file swap	[115, 116]

of the most common power systems for small electronics and therefore have been widely exploited for most wearable systems [120, 121]. Lithium-ion batteries are still the most attractive power source as they have a long and useful life cycle, low self-discharge rate, fast charge and discharge capacity, and high energy efficiency [122]. However, these conventional batteries are normally bulky and heavy, hindering its application in such type of portable device, and making it uncomfortable to the user. In addition, these conventional batteries are normally rigid, and by reducing its size, the net energy stored in the battery also decreases, thus limiting its long-term use. However, in the case of wearable sensors, a power source that charges very fast and has a long service life is required so the device can perform continuous monitoring, and so users don't have the inconvenience of having to recharge the sensor constantly [119].

Power source for wearable electrochemical devices has largely focused on three aspects of it: developing (i) energy-efficient low-power devices [123–125], (ii) compact and dense energy sources [126–128], and (iii) adaptive algorithms for smart energy saving in electronic systems [124, 129, 130]. Several approaches have been exploited so far and used for powering these wearable devices, such as usable batteries [120, 121, 131], supercapacitors [127, 128, 132], photovoltaic cells [133, 134], biofuel cells [135, 136], thermoelectric generators [137, 138], and piezoelectric/triboelectric sources [139–141]. Some studies have already shown the development of flexible energy sources, which is more compatible with on-body continuous analysis and that exhibited excellent performance for powering wearable sensors. Examples of such flexible power technology are presented in Fig. 5.

Although batteries are still the most common power source for wearable devices, supercapacitors have gain considerable attention lately as the most promising alternative for wearable applications, due to their fast charge/discharge capability, long

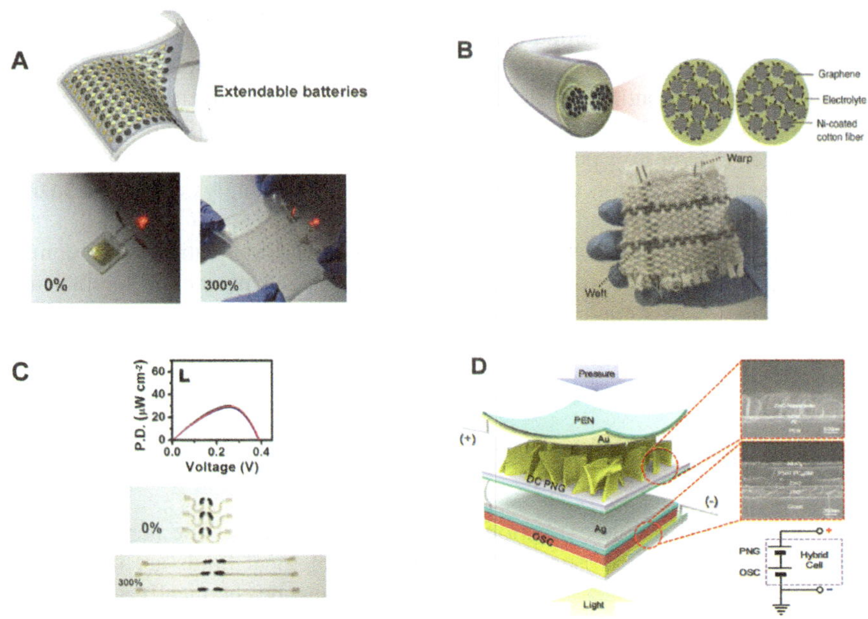

Fig. 5 Solutions to challenges based on energy sources. Pictures and diagrams illustrating **A** extendable batteries. Reproduced with permission from [120]. **B** Wearable wire supercapacitors. Reproduced with permission from [132]. **C** Highly extensible biofuel cells. Reproduced with permission from [142]. **D** A single platform containing a piezoelectric nanogenerator and photovoltaic device for harvesting energy from multiple sources. Reproduced with permission from [143]

life cycle and safety [144–146]. Wearable supercapacitors can be integrated into energy harvesting devices to store the scavenged energy [140], which can later be used to power the wearable (bio)sensor device.

Another interesting approach has emerged for powering wearable electrochemical devices using energy sources harnessed from its environment (e.g., ambient light, body movement, sweat, heat, movement, or friction), which are produced autonomously and independently of other external energy sources [119]. These self-powered systems represent an innovative step toward the development of autonomous wearable sensors, which ensure continuous energy to operate the system, and which explore multiple sources for its own energy production [126, 147]. Based on the energy harvesting mechanisms or the type of resources used, self-powered devices can be divided into: (i) self-powered photovoltaic devices based on solar energy, (ii) self-powered thermoelectric devices based on thermal energy, and (iii) integrated nanogenerator self-powered devices based on biomechanical or vibrational energy [148–150].

Biofuel cells (BFCs) are an alternative approach commonly used in self-powered wearable sensors [151–154], which uses biological molecules (e.g., metabolites) present in biofluids, such as sweat, saliva, tears, and ISF, as biofuel. However, it is important to state that this type of power source is extremely dependent on the human body physiology, and the continuous delivery of biofuel to power the device is hardly controllable. Therefore, some of the key challenges in this field of self-powered chemical sensors include continuous, stable, and prolonged power delivery. Other wearable self-powered sensors use enzymatic reactions coupled with the piezoelectric [155], triboelectric [156] or flow potential [157] generators. Wearable sensors powered by triboelectric nanogenerators (TENGs) are becoming increasingly popular due to their cheap, flexible materials and simple construction [158].

While self-powered wearable sensor developments and applications are becoming a reality, the pace of progress indicates that we will soon see many more sophisticated and fully functional examples of such technology. Figure 6 shows some examples of promising self-powered wearable sensors already presented in the literature.

3.4 Sampling

Efficient real-time sample extraction from the body is one of the most important aspects of a wearable (bio)sensor device. Some biological fluids are more easily sampled (e.g., sweat and tears), and others like ISF must be chemically and/or physically stimulated [24]. Wearable systems should have a well-adapted extraction unit capable of sampling enough volume of biofluids constantly and in a controllable manner so the effects of possible contamination and/or evaporation are minimized. Therefore, a lot of effort has been made to develop strategies to extract efficiently and continuously each one of the previously mentioned non-invasive biological fluids, coupling these strategies with various types of wearable devices [87, 162, 163].

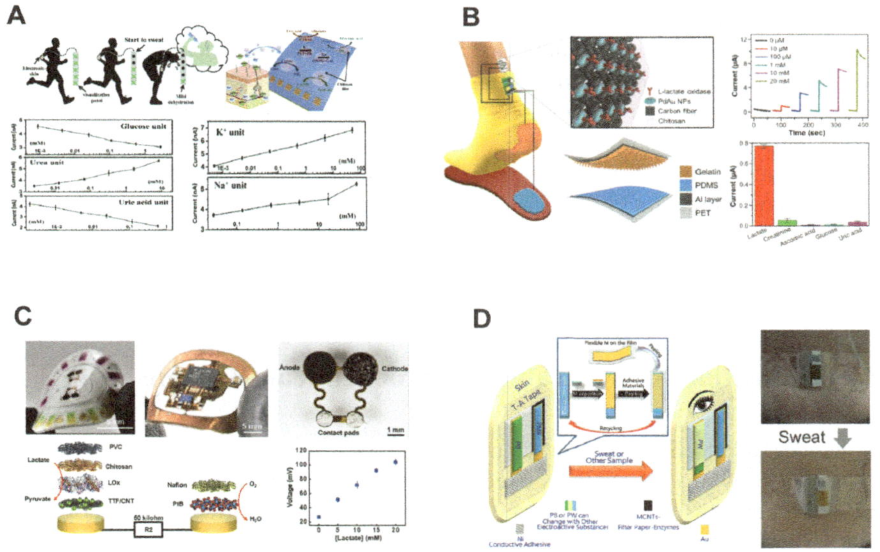

Fig. 6 Examples of the recent developments in self-powered analytical chemistry. **A** Self-powered triboelectrification-based epidermal sensors for real-time perspiration multi analysis. Reproduced with permission from [156]. **B** Self-powered triboelectrification-based lactate sensors. Reproduced with permission from [159]. **C** Epidermal self-powered BFC-based biosensors integrated with colorimetric sensors. Reproduced with permission from [160]. **D** Flexible electrochromic systems for self-powered wearable sensors. Reproduced with permission from [161]

Sweat can be naturally extracted trough exercise/physical motion or even through thermal heat, which might be interesting for some applications like fitness monitoring. However, most applications require sampling in a sedentary condition and during basic daily activities, where environment temperature and body motion can be quite variable [24, 82]. In these cases, the literature has shown a chemical stimulation of sweat by iontophoresis associated with drug-induced mechanisms. Briefly, wearable iontophoretic systems are attached to the body to deliver pilocarpine and acetylcholine through electrical current to sweat glands, inducing the biofluid extraction. Some recent works in the literature have already demonstrated the use of such biofluid extraction approach in wearable electrochemical devices [164–166].

Tears and saliva can be easily and naturally sampled with no need for extra stimulation, especially when using contact lenses-based wearable (bio)sensor and mouthguard-based (bio)sensor devices, respectively. For this reason, there are just few works mentioning some extraction approach for these biofluids in wearable settings, especially designed to minimize electrodes biofouling [167–169]. On the other hand, ISF is a biofluid that cannot be naturally secreted through the skin and therefore needs external apparatus to ensure its extraction. Ways of sampling ISF were already discussed at the beginning of this chapter, where reverse iontophoresis was presented as the most promising approach for this task [86, 170]. Furthermore, the use of microneedles has also been explored for secretion of ISF in a minimally invasive manner since it does not penetrate deep enough in the dermis and therefore does not cause pain nor discomfort to the user [171].

It is noteworthy that micro-nanofluidic systems can play an important role on controlling sample volume that reach the sensor in applications involving wearable electrochemical devices, minimizing electrode biofouling, sample leakage, evaporation, and contamination [172, 173]. Therefore, research in development of microfluidic systems coupled to sampling strategies in wearable devices remains of paramount importance.

4 Applications of Wearable Electrochemical (Bio)sensors

In the wearable technology market, miniaturized electrochemical biosensor devices have been growing exponentially over the past years, mostly related to clinical monitoring. According to the research published by Ajami et al. (2015), the field of wearable electrochemical biosensors is expected to grow by around US$70 billion by 2025, led mainly by the healthcare sector [174]. However, the market of wearable electrochemical (bio)sensors doesn't seem to be limited by one type of application; and in addition to healthcare, such devices are being developed for environmental monitoring [175], forensic analysis [28], military use and security [176]. In this topic, we are going to present a general overview of wearable biosensing applications, emphasizing electrochemical systems, for healthcare, environmental and security, including some commercially available products. Other examples of wearable biosensing applications that were not covered here can be found in some recent review articles in the literature [24, 28, 45, 75, 177].

4.1 Health Monitoring

Among the different types of biosensors with clinical application, electrochemical biosensors stand out due to their inherent system peculiarities, such as easy miniaturization, low energy consumption requirements, simplicity, fast analysis, and low manufacturing cost. These characteristics of electrochemical systems make such devices quite attractive for fabrication of flexible and wearable biosensors, as stated earlier in this chapter [94]. The potential usefulness of such devices is evident through the growing rate of recently reported work [28, 178], which have shown a large emphasis on the electrochemical field.

Table 2 contains some data published in recent literature, referring to commercial wearable electrochemical biosensors for health assessment, highlighting minimally invasive glucose monitoring devices. As stated by Kim et al. (2019) in his work, these devices still require more large-scale validation studies and regulatory approvals,

Table 2 Examples of commercial noninvasive or minimally invasive wearable electrochemical biosensors for glucose monitoring

Product	Company	Analyte sample	Wearable platform	Website
Smart contact lens	Google and Novartis	Glucose in tears	Contact lens	https://verily.com/projects/sensors/smart-lens-program/
Freestyle Libre	Abbott	Glucose in ISF	Patch	https://www.freestylelibre.us/
Dexcom G6 CGM	Dexcom	Glucose in ISF	Patch	https://www.dexcom.com/
NovioSense tear glucose sensor	NovioSense	Glucose in tears	Small stick (spiral type) placed under the lower eyelid	http://noviosense.com/

besides improved reliability, and comfort, to reach full commercialization without triggering the skepticism from the industry and/or the consumers [179].

As it can be seen, development of painless and non-invasive glucose monitoring devices is leading the advances in wearable (bio)sensing applications outside the fitness field due to the enormous interest in diabetes management [22, 180]. Although most effort toward commercialization is largely focused on glucose monitoring devices, wearable electrochemical (bio)sensors are being integrated directly into the epidermis or textile materials for monitoring a variety of relevant chemical species (e.g., metabolites and electrolytes) [181]. There are review articles in the literature discussing several examples of wearable electrochemical (bio)sensors that have been developed so far, especially for health assessment [74, 179, 182–184].

In 2020, the world faced (and is still facing it the time we are publishing this book chapter) a devastating pandemic scenario caused by the new coronavirus, named SARS-Cov-2.[3] The COVID-19 pandemic (as it is called) brought a lot of attention and efforts of scientists all over the globe to develop new drugs [185], treatments [186], vaccines [187], rapid tests [188], and monitoring devices [189]. In this scenario, wearable devices may play an important role on virus spread monitoring, besides providing relevant real-time data of patients already contaminated during their treatment, even outside the hospital or in remote regions. Un and co-workers (2021) have reported an observational study exploring the potential of using wearable biosensors for managing COVID-19 patients [189]. In this case, they coupled the wearable device to a machine learning-based remote monitoring platform, which allowed automated assessment of clinical deterioration. Ates et al. (2021) has also published an article on wearable electronic (bio)sensors for COVID-19 management, highlighting that such devices can be used in the early detection of asymptomatic and pre-symptomatic cases of COVID-19 through continuous monitoring of physiological signals [190]. The authors also added some examples in the literature of emerging

[3] Severe Acute Respiratory Syndrome Coronavirus 2 (SARS-Cov-2).

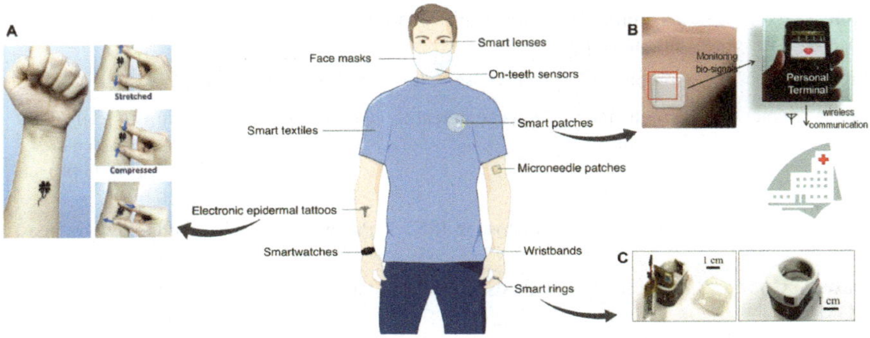

Fig. 7 Possibilities of future wearable electrochemical (bio)sensor devices for healthcare. Adapted with permission from [190]. Images showing **A** four-leaf clover electronic tattoo attached to the forearm (reproduced with permission from [191]); **B** illustration of dry adhesive as biomedical bandage in a real-time diagnosis system (reproduced with permission from [192]); and **C** self-sustainable smart ring for long-term monitoring of blood oxygenation (reproduced with permission from [193])

low-cost wearable electrochemical (bio)sensors that could extend the capabilities of such devices, as shown in Fig. 7.

It has been demonstrated that wearable (bio)sensors can be used to measure several relevant health parameters, such as heart rate, blood pressure, respiration rate, among various biomarkers. Therefore, understanding that wearable technology has effective tools for timely prevention, diagnosis, treatment, and control of diseases becomes crucial, especially during disease outbreaks [174]. It is noteworthy that despite the focus of wearable electrochemical devices for clinical application, the development of such devices has been gaining ground in other areas such as forensic and environmental monitoring, which will be discussed next.

4.2 Security and Environmental Monitoring

Although wearable (bio)sensors are typically associated with clinical applications, such devices may effectively be employed to monitor the environment around the user, for instance. Detection of drugs, explosives, adulterations, and toxic compounds are examples of applications where wearable electrochemical (bio)sensor are mostly welcome [28]. In the literature, there are considerable demands for research regarding the implementation of wearable sensors for tracking the local environment to identify potential causes of criminal activity in toxicological and forensic investigations. Alcohol consumption is common throughout society, so monitoring alcohol abuse in an effective real-time detection also represents substantial impact on personal and automotive safety [28].

At the forensic level, Teymourian and colleagues (2020) published a review article on the progress of wearable electrochemical sensors for monitoring therapeutic drugs and drugs of abuse [194]. The authors report that they have evidenced the increasing progress in the use of wearable electrochemical devices for field screening of drugs of abuse for various forensic applications. They further emphasize that the main recent efforts have been directed toward combining wearable devices with different electrochemical techniques in the development of wearable platforms for drug monitoring. Ferreira et al. (2019) have dedicated a whole topic of his review article to discuss wearable electrochemical (bio)sensor in forensic application, highlighting the increasing relevance of this type of devices in such application [28]. Nowadays, there are interesting promising wearable technologies for forensic applications, such as a ring-shaped wearable electrochemical sensor for rapid and simultaneous identification of salivary tetrahydrocannabinol and alcohol, and a tattoo-based skin-integrated wearable sensor also to detect alcohol. Figure 8 shows representative examples of wearable electrochemical sensors for drug analysis.

Detection of explosives, especially 2,4,6-trinitrotoluene (TNT), attract worldwide interest because of the threats to public safety, as well as to human health, as they have become pollutants in water and other environmental samples [197]. The inherent redox activity of widely used nitroaromatic explosives like TNT makes them ideal candidates for electrochemical detection [45]. In this regard, Zhang et al. developed an impedimetric biosensor for monitoring TNT explosive samples using a portable

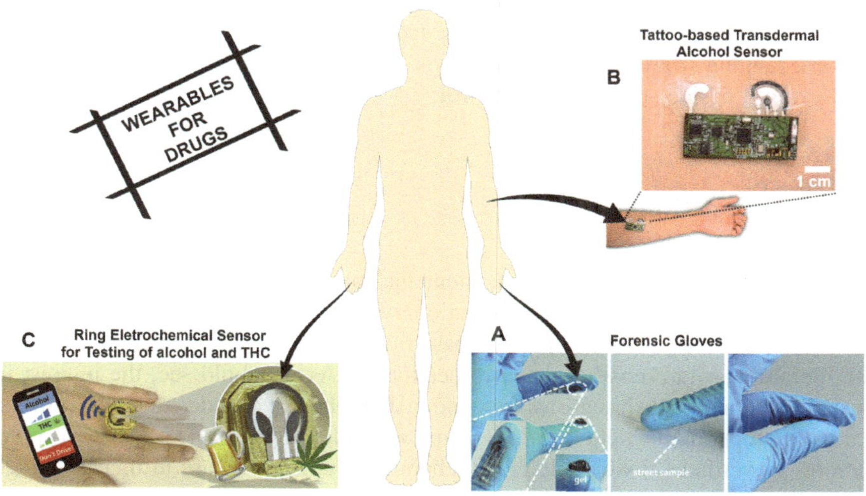

Fig. 8 Representative electrochemical wearables for drug analysis. **A** External drug screening: wearable glove sensor for analysis of suspicious powder samples, showing screen-printed electrode onto the flexible finger, and a conductive gel immobilized upon the thumb. Adapted with permission from [195]. **B** Sweat monitoring: image of a skin-worn alcohol iontophoretic-sensing tattoo device with integrated flexible electronic board. Adapted with permission from [166]. **C** Saliva monitoring: ring-based THC/alcohol sensor. Adapted with permission from [196]

hand-held electronic device and a smartphone [197]. The electrochemical (bio)sensor consisted of screen-printed electrodes functionalized with TNT-specific peptides, and the responses were indirectly obtained through inhibited electron transfer in the presence of the analyte. Finally, the data is collected by a portable device and sent to a smartphone via Bluetooth. Although this is not a wearable device, the idea behind this work can be easily adapted for wearable applications.

Wearable glove-based sensors is another promising technology for monitoring environmental problems by bringing chemical analyzes directly to the user's fingertips [28]. The "slide, scan, detect and alert" strategy is quite attractive and raise the possibility of performing real-time detection and screening of target analytes in emergency situations. In this context, Mishra and co-workers (2017) proposed a flexible electrochemical biosensor employing a elastic-printed electrode system that resists mechanical deformations and detects different organophosphate (OP) nerve-agent compounds [30]. The electrochemical detection system was processed when the thumb encountered the sensor (index) finger, and a semi-solid conducting gel matrix containing organophosphate hydrolase (OPH). Then, the OP compound present in the sample was hydrolyzed generating *p*-nitrophenol as electroactive molecule, which that was detected electrochemically, with the result wirelessly transmitted to a smartphone. Zhao et al. (2020) proposed a similar approach for detecting OP pesticides, but with the wearable flexible electrochemical biosensor directly attached to the agricultural product surface for environmental monitoring, as it can be seen in Fig. 9 [59].

These examples of wearable miniaturized electrochemical (bio)sensors shown here highlights the possibilities of using such devices for a broad range of applications, ranging from health and fitness monitoring to environmental and forensic applications, besides food quality control.

5 Considerations

In this chapter, we have discussed about the main components and approaches, besides some key aspects for fabrication of reliable wearable electrochemical (bio)sensors. Those discussions were based on works that have been published in the literature so far, mostly related to healthcare. As we could see, the interest in wearable technologies has grown tremendously in the last decade, and it will still be a trending topic in this field for many years, with a lot of prospects for several applications. Nevertheless, wearable electrochemical (bio)sensors are in its early stage of development, and many challenges need to be overcome before we can see these devices reaching the market.

Some of the main challenges faced in wearable electrochemical (bio)sensors field, which may act as seeds for new research projects, include: (i) mass production; (ii) integration of the miniaturized electrochemical system with wireless electronics; (iii) need for fluidic systems and actuators coupled to the wearable platform (for automation and complex sample preparation); (iv) continuous monitoring; (v)

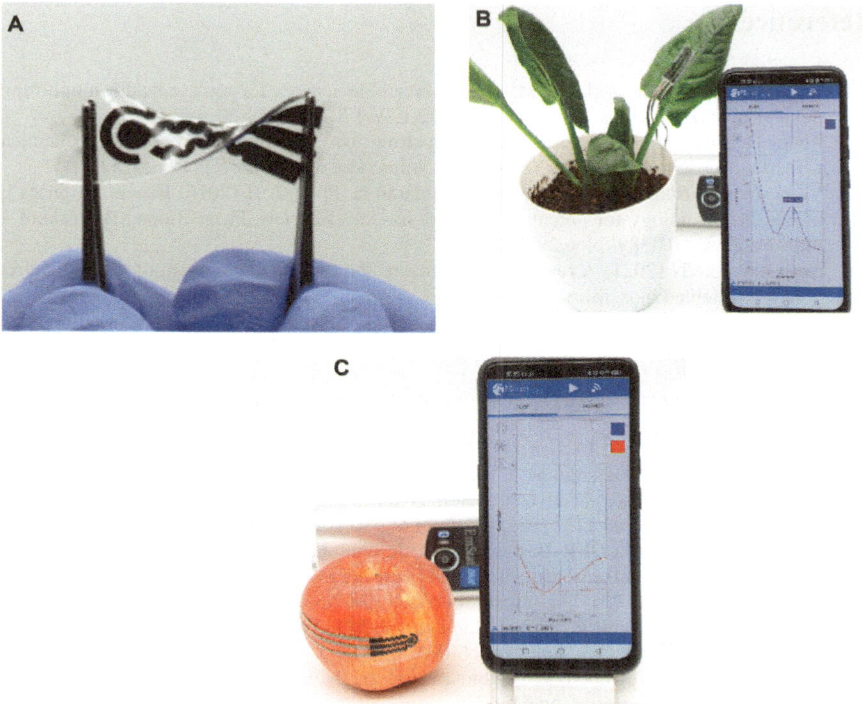

Fig. 9 Digital photo of **A** the electrode under twisting; **B**, **C** in situ OP analysis on the surfaces of different agricultural products using a pocketsize potentiostat and a smartphone as readout system. Reproduced with permission from [59]

internal analytical curves with built-in automatic correction based on changes in the environment (e.g., pH, temperature, and mechanical stress); (vi) mechanisms to avoid electrode biofouling; (vii) new non-invasive sampling systems; (viii) artificial organs for proper validation tests; (ix) prolonged electrochemical sensing; and (x) compact, flexible, high-energy batteries to power everything up.

Even with all these challenges to face, electrochemistry remains on top for wearable applications, with easy miniaturization and integration with electronic components, besides its outstanding sensing properties. After everything we discussed in this chapter, we can say that electrochemistry may be leading a new revolution on wearable sensors not just for fitness monitoring, but for healthcare, security, forensic, environmental, and food quality control as well. For this to happen, a multidisciplinary approach (e.g., professionals of chemistry, physics, biology, engineering, computer scientists working together) is necessary to fill the gaps that hinder these devices to reach real applications and commercialization.

References

1. Majumder S, Mondal T, Deen MJ (2017) Wearable sensors for remote health monitoring. Sensors (Switzerland) 17. https://doi.org/10.3390/s17010130
2. Zhang D, Liu Q (2016) Biosensors and bioelectronics on smartphone for portable biochemical detection. Biosens Bioelectron 75:273–284. https://doi.org/10.1016/j.bios.2015.08.037
3. Vigneshvar S, Sudhakumari CC, Senthilkumaran B, Prakash H (2016) Recent advances in biosensor technology for potential applications—an overview. Front Bioeng Biotechnol 4. https://doi.org/10.3389/fbioe.2016.00011
4. Naresh V, Lee N (2021) A review on biosensors and recent development of nanostructured materials-enabled biosensors
5. Patel S, Nanda R, Sahoo S, Mohapatra E (2016) Biosensors in health care: the milestones achieved in their development towards lab-on-chip-analysis. Biochem Res Int 2016:3130469. https://doi.org/10.1155/2016/3130469
6. Chen J, Wu W, Zeng L (2014) A universal glucometer-based biosensor for portable and quantitative detection of transcription factors. Anal Methods 6:4840–4844. https://doi.org/10.1039/C4AY00323C
7. Maduraiveeran G, Sasidharan M, Ganesan V (2018) Electrochemical sensor and biosensor platforms based on advanced nanomaterials for biological and biomedical applications. Biosens Bioelectron 103:113–129. https://doi.org/10.1016/j.bios.2017.12.031
8. da Silva ETSG, Souto DEP, Barragan JTC, de Giarola FJ, de Moraes ACM, Kubota LT (2017) Electrochemical biosensors in point-of-care devices: recent advances and future trends. ChemElectroChem 4:778–794. https://doi.org/10.1002/celc.201600758
9. Grieshaber D, MacKenzie R, Vörös J, Reimhult E (2008) Electrochemical biosensors—sensor principles and architectures. Sensors 8:1400
10. Sadeghi SJ (2013) Amperometric biosensors. In: Roberts GCK (ed) Encyclopedia of biophysics. Springer, Berlin, pp 61–67
11. Tarasov A, Gray DW, Tsai M-Y, Shields N, Montrose A, Creedon N, Lovera P, O'Riordan A, Mooney MH, Vogel EM (2016) A potentiometric biosensor for rapid on-site disease diagnostics. Biosens Bioelectron 79:669–678. https://doi.org/10.1016/j.bios.2015.12.086
12. Guan J-G, Miao Y-Q, Zhang Q-J (2004) Impedimetric biosensors. J Biosci Bioeng 97:219–226. https://doi.org/10.1016/S1389-1723(04)70195-4
13. Hammond JL, Formisano N, Estrela P, Carrara S, Tkac J (2016) Electrochemical biosensors and nanobiosensors. Essays Biochem 60:69–80. https://doi.org/10.1042/EBC20150008
14. Liu B, Du D, Hua X, Yu X-Y, Lin Y (2014) Paper-based electrochemical biosensors: from test strips to paper-based microfluidics. Electroanalysis 26:1214–1223. https://doi.org/10.1002/elan.201400036
15. Darwish NT, Sekaran SD, Khor SM (2017) Point-of-care tests: a review of advances in the emerging diagnostic tools for dengue virus infection. Sensors Actuators B Chem. https://doi.org/10.1016/j.snb.2017.09.159
16. Clark LC Jr, Lyons C (1962) Electrode systems for continuous monitoring in cardiovascular surgery. Ann N Y Acad Sci 102:29–45. https://doi.org/10.1111/j.1749-6632.1962.tb13623.x
17. Amalfitano E, Karlikow M, Norouzi M, Jaenes K, Cicek S, Masum F, Sadat Mousavi P, Guo Y, Tang L, Sydor A, Ma D, Pearson JD, Trcka D, Pinette M, Ambagala A, Babiuk S, Pickering B, Wrana J, Bremner R, Mazzulli T, Sinton D, Brumell JH, Green AA, Pardee K (2021) A glucose meter interface for point-of-care gene circuit-based diagnostics. Nat Commun 12:724. https://doi.org/10.1038/s41467-020-20639-6
18. Lan T, Zhang J, Lu Y (2016) Transforming the blood glucose meter into a general healthcare meter for in vitro diagnostics in mobile health. Biotechnol Adv 34:331–341. https://doi.org/10.1016/j.biotechadv.2016.03.002
19. Abardía-Serrano C, Miranda-Castro R, de-los-Santos-Álvarez N, Lobo-Castañón MJ (2020) New uses for the personal glucose meter: detection of nucleic acid biomarkers for prostate cancer screening

20. Li X, Huang X, Mo J, Wang H, Huang Q, Yang C, Zhang T, Chen H-J, Hang T, Liu F, Jiang L, Wu Q, Li H, Hu N, Xie X (2021) A fully integrated closed-loop system based on mesoporous microneedles-iontophoresis for diabetes treatment. Adv Sci 8:2100827. https://doi.org/10.1002/advs.202100827
21. Ji D, Shi Z, Liu Z, Low SS, Zhu J, Zhang T, Chen Z, Yu X, Lu Y, Lu D, Liu Q (2020) Smartphone-based square wave voltammetry system with screen-printed graphene electrodes for norepinephrine detection. Smart Mater Med 1:1–9. https://doi.org/10.1016/j.smaim.2020.02.001
22. Zhao J, Lin Y, Wu J, Nyein HYY, Bariya M, Tai LC, Chao M, Ji W, Zhang G, Fan Z, Javey A (2019) A fully integrated and self-powered smartwatch for continuous sweat glucose monitoring. ACS Sensors. 4:1925–1933. https://doi.org/10.1021/acssensors.9b00891
23. Nag A, Mukhopadhyay SC, Kosel J (2017) Wearable flexible sensors: a review. IEEE Sens J 17:3949–3960. https://doi.org/10.1109/JSEN.2017.2705700
24. Min J, Sempionatto JR, Teymourian H, Wang J, Gao W (2021) Wearable electrochemical biosensors in North America. Biosens Bioelectron 172:112750. https://doi.org/10.1016/j.bios.2020.112750
25. Tierney MJ, Tamada JA, Potts RO, Eastman RC, Pitzer K, Ackerman NR, Fermi SJ (2000) The GlucoWatch® biographer: a frequent, automatic and noninvasive glucose monitor. Ann Med 32:632–641. https://doi.org/10.3109/07853890009002034
26. Tierney MJ, Tamada JA, Potts RO, Jovanovic L, Garg S (2001) Clinical evaluation of the GlucoWatch® biographer: a continual, non-invasive glucose monitor for patients with diabetes. Biosens Bioelectron 16:621–629. https://doi.org/10.1016/S0956-5663(01)00189-0
27. Alves T, Deroco P, Wachholz Junior D, Vidotto L, Kubota L (2021) Wireless wearable electrochemical sensors: a review. Brazilian J Anal Chem 8:1–29. https://doi.org/10.30744/brjac.2179-3425.rv-62-2020
28. Ferreira PC, Ataíde VN, Silva Chagas CL, Angnes L, Tomazelli Coltro WK, Longo Cesar Paixão TR, de Araujo WR (2019) Wearable electrochemical sensors for forensic and clinical applications. TrAC Trends Anal Chem 119. https://doi.org/10.1016/j.trac.2019.115622
29. Pandian PS, Mohanavelu K, Safeer KP, Kotresh TM, Shakunthala DT, Gopal P, Padaki VC (2008) Smart vest: wearable multi-parameter remote physiological monitoring system. Med Eng Phys 30. https://doi.org/10.1016/j.medengphy.2007.05.014
30. Mishra RK, Hubble LJ, Martín A, Kumar R, Barfidokht A, Kim J, Musameh MM, Kyratzis IL, Wang J (2017) Wearable flexible and stretchable glove biosensor for on-site detection of organophosphorus chemical threats. ACS Sensors 2:553–561. https://doi.org/10.1021/acssensors.7b00051
31. Fabietti PG, Massi Benedetti M, Bronzo F, Reboldi GP, Sarti E, Brunetti P (1991) Wearable system for acquisition, processing and storage of the signal from amperometric glucose sensors. Int J Artif Organs 14:175–178
32. Zhang W, Wang R, Luo F, Wang P, Lin Z (2020) Miniaturized electrochemical sensors and their point-of-care applications. Chin Chem Lett 31:589–600. https://doi.org/10.1016/j.cclet.2019.09.022
33. Silva ETSG, Santhiago M, Barragan JTC, Kubota LT (2014) Construction of a new versatile point-of-care testing device with electrochemical detection employing paper as a microfluidic platform. Anal Methods 6:6133–6136. https://doi.org/10.1039/C4AY00986J
34. Rowe AA, Bonham AJ, White RJ, Zimmer MP, Yadgar RJ, Hobza TM, Honea JW, Ben-Yaacov I, Plaxco KW (2011) CheapStat: an open-source, "do-it-yourself" potentiostat for analytical and educational applications. PLoS ONE 6:e23783
35. Steinberg MD, Kassal P, Kereković I, Steinberg IM (2015) A wireless potentiostat for mobile chemical sensing and biosensing. Talanta 143:178–183. https://doi.org/10.1016/j.talanta.2015.05.028
36. Delaney JL, Doeven EH, Harsant AJ, Hogan CF (2013) Use of a mobile phone for potentiostatic control with low cost paper-based microfluidic sensors. Anal Chim Acta 790:56–60. https://doi.org/10.1016/j.aca.2013.06.005

37. Stradiotto NR, Yamanaka H, Zanoni MVB (2003) Electrochemical sensors: a powerful tool in analytical chemistry. J Braz Chem Soc 14:159–173. https://doi.org/10.1590/S0103-50532003000200003
38. Abdulbari HA, Basheer EAM (2017) Electrochemical biosensors: electrode development, materials, design, and fabrication. ChemBioEng Rev 4:92–105. https://doi.org/10.1002/cben.201600009
39. da Silva ETSG, Alves TMR, Kubota LT (2018) Direct toner printing: a versatile technology for easy fabrication of flexible miniaturized electrodes. Electroanalysis 30:345–352. https://doi.org/10.1002/elan.201700717
40. da Silva ET, Miserere S, Kubota LT, Merkoci A (2014) Simple on-plastic/paper inkjet-printed solid-state Ag/AgCl pseudoreference electrode. Anal Chem 86:10531–10534. https://doi.org/10.1021/ac503029q
41. Wu W (2019) Stretchable electronics: functional materials, fabrication strategies and applications. Sci Technol Adv Mater 20:187–224. https://doi.org/10.1080/14686996.2018.1549460
42. Fan JA, Yeo W-H, Su Y, Hattori Y, Lee W, Jung S-Y, Zhang Y, Liu Z, Cheng H, Falgout L, Bajema M, Coleman T, Gregoire D, Larsen RJ, Huang Y, Rogers JA (2014) Fractal design concepts for stretchable electronics. Nat Commun 5:3266. https://doi.org/10.1038/ncomms4266
43. Taleat Z, Khoshroo A, Mazloum-Ardakani M (2014) Screen-printed electrodes for biosensing: a review (2008–2013). Microchim Acta 181:865–891. https://doi.org/10.1007/s00604-014-1181-1
44. Yáñez-Sedeño P, Campuzano S, Pingarrón JM (2020) Screen-printed electrodes: promising paper and wearable transducers for (bio)sensing. Biosensors 10. https://doi.org/10.3390/BIOS10070076
45. Windmiller JR, Wang J (2013) Wearable electrochemical sensors and biosensors: a review. Electroanalysis 25:29–46. https://doi.org/10.1002/elan.201200349
46. Metters JP, Kadara RO, Banks CE (2011) New directions in screen printed electroanalytical sensors: an overview of recent developments. Analyst 136:1067–1076. https://doi.org/10.1039/C0AN00894J
47. Bariya M, Shahpar Z, Park H, Sun J, Jung Y, Gao W, Nyein HYY, Liaw TS, Tai LC, Ngo QP, Chao M, Zhao Y, Hettick M, Cho G, Javey A (2018) Roll-to-roll gravure printed electrochemical sensors for wearable and medical devices. ACS Nano 12:6978–6987. https://doi.org/10.1021/acsnano.8b02505
48. Li J, Rossignol F, Macdonald J (2015) Inkjet printing for biosensor fabrication: combining chemistry and technology for advanced manufacturing. Lab Chip 15:2538–2558. https://doi.org/10.1039/C5LC00235D
49. Gao M, Li L, Song Y (2017) Inkjet printing wearable electronic devices. J Mater Chem C 5:2971–2993. https://doi.org/10.1039/C7TC00038C
50. Al-Halhouli A, Qitouqa H, Alashqar A, Abu-Khalaf J (2018) Inkjet printing for the fabrication of flexible/stretchable wearable electronic devices and sensors. Sens Rev 38:438–452. https://doi.org/10.1108/SR-07-2017-0126
51. Maejima K, Tomikawa S, Suzuki K, Citterio D (2013) Inkjet printing: an integrated and green chemical approach to microfluidic paper-based analytical devices. Rsc Adv 3:9258–9263. https://doi.org/10.1039/c3ra40828k
52. Silva AL, da Salvador GMS, Castro SVF, Carvalho NMF, Munoz RAA (2021) A 3D printer guide for the development and application of electrochemical cells and devices. https://www.frontiersin.org/article/10.3389/fchem.2021.684256
53. Katseli V, Thomaidis N, Economou A, Kokkinos C (2020) Miniature 3D-printed integrated electrochemical cell for trace voltammetric Hg(II) determination. Sensors Actuators B Chem 308:127715. https://doi.org/10.1016/j.snb.2020.127715
54. Padash M, Carrara S (2020) A 3D printed wearable device for sweat analysis. In: 2020 IEEE international symposium on medical measurements and applications (MeMeA), pp 1–5

55. Katseli V, Economou A, Kokkinos C (2021) Smartphone-addressable 3D-printed electrochemical ring for nonenzymatic self-monitoring of glucose in human sweat. Anal Chem 93:3331–3336. https://doi.org/10.1021/acs.analchem.0c05057
56. Lin J, Peng Z, Liu Y, Ruiz-Zepeda F, Ye R, Samuel ELG, Yacaman MJ, Yakobson BI, Tour JM (2014) Laser-induced porous graphene films from commercial polymers. Nat Commun 5:5714. https://doi.org/10.1038/ncomms6714
57. Kumar R, Joanni E, Singh RK, da Silva ETSG, Savu R, Kubota LT, Moshkalev SA (2017) Direct laser writing of micro-supercapacitors on thick graphite oxide films and their electrochemical properties in different liquid inorganic electrolytes. J Colloid Interface Sci 507:271–278. https://doi.org/10.1016/j.jcis.2017.08.005
58. Ye R, James DK, Tour JM (2019) Laser-induced graphene: from discovery to translation. Adv Mater 31:e1803621. https://doi.org/10.1002/adma.201803621
59. Zhao F, He J, Li X, Bai Y, Ying Y, Ping J (2020) Smart plant-wearable biosensor for in-situ pesticide analysis. Biosens Bioelectron 170:112636. https://doi.org/10.1016/j.bios.2020.112636
60. Yang Y, Song Y, Bo X, Min J, Pak OS, Zhu L, Wang M, Tu J, Kogan A, Zhang H, Hsiai TK, Li Z, Gao W (2020) A laser-engraved wearable sensor for sensitive detection of uric acid and tyrosine in sweat. Nat Biotechnol 38:217–224. https://doi.org/10.1038/s41587-019-0321-x
61. Zhang X, Guo S, Han Y, Li J, Wang E (2017) Beyond conventional patterns: new electrochemical lithography with high precision for patterned film materials and wearable sensors. Anal Chem 89:2569–2574. https://doi.org/10.1021/acs.analchem.6b04816
62. Ma D, Chon S, Cho S, Lee Y, Yoo M, Kim D, Lee DY, Lim JK (2020) A novel photolithographic method for fabrication of flexible micro-patterned glucose sensors. J Electroanal Chem 876:114720. https://doi.org/10.1016/j.jelechem.2020.114720
63. Hyun WJ, Secor EB, Hersam MC, Frisbie CD, Francis LF (2015) High-resolution patterning of graphene by screen printing with a silicon stencil for highly flexible printed electronics. Adv Mater 27:109–115. https://doi.org/10.1002/adma.201404133
64. Bae S, Kim H, Lee Y, Xu X, Park J-S, Zheng Y, Balakrishnan J, Lei T, Ri Kim H, Song YI, Kim Y-J, Kim KS, Özyilmaz B, Ahn J-H, Hong BH, Iijima S (2010) Roll-to-roll production of 30-inch graphene films for transparent electrodes. Nat Nanotechnol 5:574–578. https://doi.org/10.1038/nnano.2010.132
65. Jabari E, Ahmed F, Liravi F, Secor EB, Lin L, Toyserkani E (2019) 2D printing of graphene: a review. 2D Mater 6:42004. https://doi.org/10.1088/2053-1583/ab29b2
66. Lu Z, Lin X, Zhang J, Dai W, Liu B, Mo G, Ye J, Ye J (2019) Ionic liquid/poly-l-cysteine composite deposited on flexible and hierarchical porous laser-engraved graphene electrode for high-performance electrochemical analysis of lead ion. Electrochim Acta 295:514–523. https://doi.org/10.1016/j.electacta.2018.10.176
67. Verma A, Jennings J, Johnson RD, Weber MH, Lynn KG (2013) Fabrication of 3D charged particle trap using through-silicon vias etched by deep reactive ion etching. J Vac Sci Technol B 31:32001. https://doi.org/10.1116/1.4799662
68. Li Z, Liu H, He X, Xu F, Li F (2018) Pen-on-paper strategies for point-of-care testing of human health. TrAC Trends Anal Chem 108:50–64. https://doi.org/10.1016/j.trac.2018.08.010
69. Kurra N, Kulkarni GU (2013) Pencil-on-paper: electronic devices. Lab Chip 13:2866–2873. https://doi.org/10.1039/c3lc50406a
70. Li W, Qian D, Li Y, Bao N, Gu H, Yu C (2016) Fully-drawn pencil-on-paper sensors for electroanalysis of dopamine. J Electroanal Chem 769:72–79. https://doi.org/10.1016/j.jelechem.2016.03.027
71. Dossi N, Toniolo R, Pizzariello A, Impellizzieri F, Piccin E, Bontempelli G (2013) Pencil-drawn paper supported electrodes as simple electrochemical detectors for paper-based fluidic devices. Electrophoresis 34:2085–2091. https://doi.org/10.1002/elps.201200425
72. Chen Y, Liu X, Ma Y, Chen Y, Lu B, Feng X (2021) Interfacial liquid film transfer printing of versatile flexible electronic devices with high yield ratio. Adv Mater Interfaces 8:2100287. https://doi.org/10.1002/admi.202100287

73. Windmiller JR, Bandodkar AJ, Parkhomovsky S, Wang J (2012) Stamp transfer electrodes for electrochemical sensing on non-planar and oversized surfaces. Analyst 137:1570–1575. https://doi.org/10.1039/C2AN35041F
74. Salim A, Lim S (2019) Recent advances in noninvasive flexible and wearable wireless biosensors. Biosens Bioelectron 141:111422. https://doi.org/10.1016/j.bios.2019.111422
75. Liu H, Zhao C (2020) Wearable electrochemical sensors for noninvasive monitoring of health—a perspective
76. Foster CW, Metters JP, Banks CE (2013) Ultra flexible paper based electrochemical sensors: effect of mechanical contortion upon electrochemical performance. Electroanalysis 25:2275–2282. https://doi.org/10.1002/elan.201300274
77. Zhu Z, Li R, Pan T (2018) Imperceptible epidermal–iontronic interface for wearable sensing. Adv Mater 30:1705122. https://doi.org/10.1002/adma.201705122
78. Zhao Y, Zhai Q, Dong D, An T, Gong S, Shi Q, Cheng W (2019) Highly stretchable and strain-insensitive fiber-based wearable electrochemical biosensor to monitor glucose in the sweat. Anal Chem 91:6569–6576. https://doi.org/10.1021/acs.analchem.9b00152
79. Yang X, Cheng H (2020) Recent developments of flexible and stretchable electrochemical biosensors
80. Zhao Y, Huang X (2017) Mechanisms and materials of flexible and stretchable skin sensors
81. Harito C, Utari L, Putra BR, Yuliarto B, Purwanto S, Zaidi SZJ, Bavykin DV, Marken F, Walsh FC (2020) Review—the development of wearable polymer-based sensors: perspectives. J Electrochem Soc 167:37566. https://doi.org/10.1149/1945-7111/ab697c
82. Mohan AMV, Rajendran V, Mishra RK, Jayaraman M (2020) Recent advances and perspectives in sweat based wearable electrochemical sensors. TrAC Trends Anal Chem 131:116024. https://doi.org/10.1016/j.trac.2020.116024
83. Zardetto V, Brown TM, Reale A, Di Carlo A (2011) Substrates for flexible electronics: a practical investigation on the electrical, film flexibility, optical, temperature, and solvent resistance properties. J Polym Sci Part B Polym Phys 49:638–648. https://doi.org/10.1002/polb.22227
84. Gao W, Ota H, Kiriya D, Takei K, Javey A (2019) Flexible electronics toward wearable sensing. Acc Chem Res 52:523–533. https://doi.org/10.1021/acs.accounts.8b00500
85. Madden J, O'Mahony C, Thompson M, O'Riordan A, Galvin P (2020) Biosensing in dermal interstitial fluid using microneedle based electrochemical devices. Sens Bio-Sens Res 29:100348. https://doi.org/10.1016/j.sbsr.2020.100348
86. Kim J, Sempionatto JR, Imani S, Hartel MC, Barfidokht A, Tang G, Campbell AS, Mercier PP, Wang J (2018) Simultaneous monitoring of sweat and interstitial fluid using a single wearable biosensor platform. Adv Sci 5:1800880. https://doi.org/10.1002/advs.201800880
87. Liu C, Xu T, Wang D, Zhang X (2020) The role of sampling in wearable sweat sensors. Talanta 212:120801. https://doi.org/10.1016/j.talanta.2020.120801
88. Li H, Xiao Y, Jin L (2020) Recent development of wearable electrochemical sensors for sweat analysis. Int J Electrochem Sci 15:11997–12006. https://doi.org/10.20964/2020.12.22
89. Kim J, Imani S, de Araujo WR, Warchall J, Valdés-Ramírez G, Paixão TRLC, Mercier PP, Wang J (2015) Wearable salivary uric acid mouthguard biosensor with integrated wireless electronics. Biosens Bioelectron 74:1061–1068. https://doi.org/10.1016/j.bios.2015.07.039
90. Senior M (2014) Novartis signs up for Google smart lens. Nat Biotechnol 32:856. https://doi.org/10.1038/nbt0914-856
91. Hee KD, Su-Kyoung K, Jahyun K, Geon-Hui L, Cheonhoo J, Won MJ, Ho MB, Jae LK, Ehsan K, Choun-Ki J, Sangbaie S, Jae-Yoon S, David M, Hyun YS, Zhenan B, Kwang HS (2021) Wireless smart contact lens for diabetic diagnosis and therapy. Sci Adv 6:eaba3252. https://doi.org/10.1126/sciadv.aba3252
92. Bandodkar AJ, Jia WZ, Yardimci C, Wang X, Ramirez J, Wang J (2015) Tattoo-based noninvasive glucose monitoring: a proof-of-concept study. Anal Chem 87:394–398. https://doi.org/10.1021/ac504300n
93. Matzeu G, Florea L, Diamond D (2015) Advances in wearable chemical sensor design for monitoring biological fluids. Sensors Actuators B Chem 211:403–418. https://doi.org/10.1016/j.snb.2015.01.077

94. Heikenfeld J, Jajack A, Rogers J, Gutruf P, Tian L, Pan T, Li R, Khine M, Kim J, Wang J, Kim J (2018) Wearable sensors: modalities, challenges, and prospects. Lab Chip 18:217–248. https://doi.org/10.1039/c7lc00914c
95. Shrivastava S, Trung TQ, Lee NE (2020) Recent progress, challenges, and prospects of fully integrated mobile and wearable point-of-care testing systems for self-testing. Chem Soc Rev 49:1812–1866. https://doi.org/10.1039/c9cs00319c
96. Trevisan MG, Poppi RJ (2006) Química Analítica de Processos. Quim Nova 29:1065–1071. https://doi.org/10.1590/s0100-40422006000500029
97. Parrilla M, Cuartero M, Crespo GA (2019) Wearable potentiometric ion sensors. TrAC Trends Anal Chem 110:303–320. https://doi.org/10.1016/j.trac.2018.11.024
98. Zhou L, Li X, Zhu B, Su B (2021) An overview of antifouling strategies for electrochemical analysis. Electroanalysis. https://doi.org/10.1002/elan.202100406
99. Hanssen BL, Siraj S, Wong DKY (2016) Recent strategies to minimise fouling in electrochemical detection systems. Rev Anal Chem 35:1–28. https://doi.org/10.1515/revac-2015-0008
100. Mugo SM, Alberkant J, Bernstein N, Zenkina OV (2021) Flexible electrochemical aptasensor for cortisol detection in human sweat. Anal Methods 13:4169–4173. https://doi.org/10.1039/D1AY01233A
101. Zhu X, Ju Y, Chen J, Liu D, Liu H (2018) Nonenzymatic wearable sensor for electrochemical analysis of perspiration glucose. ACS Sensors. 3:1135–1141. https://doi.org/10.1021/acssensors.8b00168
102. Bahadır EB, Sezgintürk MK (2016) A review on impedimetric biosensors. Artif Cells Nanomedicine Biotechnol 44:248–262. https://doi.org/10.3109/21691401.2014.942456
103. Randviir EP, Banks CE (2013) Electrochemical impedance spectroscopy: an overview of bioanalytical applications. Anal Methods 5:1098–1115. https://doi.org/10.1039/C3AY26476A
104. Nah JS, Barman SC, Zahed MA, Sharifuzzaman M, Yoon H, Park C, Yoon S, Zhang S, Park JY (2021) A wearable microfluidics-integrated impedimetric immunosensor based on Ti3C2Tx MXene incorporated laser-burned graphene for noninvasive sweat cortisol detection. Sensors Actuators B Chem. 329:129206. https://doi.org/10.1016/j.snb.2020.129206
105. Zhang T, Lu J, Hu F, Hao Q (2014) Bluetooth low energy for wearable sensor-based healthcare systems. In: 2014 IEEE healthcare innovation conference (HIC), pp 251–254
106. Gao W, Emaminejad S, Nyein HYY, Challa S, Chen K, Peck A, Fahad HM, Ota H, Shiraki H, Kiriya D, Lien D-H, Brooks GA, Davis RW, Javey A (2016) Fully integrated wearable sensor arrays for multiplexed in situ perspiration analysis. Nature 529:509–514. https://doi.org/10.1038/nature16521. http://www.nature.com/nature/journal/v529/n7587/abs/nature16521.html#supplementary-information
107. Lorwongtragool P, Sowade E, Watthanawisuth N, Baumann RR, Kerdcharoen T (2014) A novel wearable electronic nose for healthcare based on flexible printed chemical sensor array. Sensors (Switzerland) 14:19700–19712. https://doi.org/10.3390/s141019700
108. Kassal P, Steinberg MD, Steinberg IM (2018) Wireless chemical sensors and biosensors: a review. Sensors Actuators, B Chem 266:228–245. https://doi.org/10.1016/j.snb.2018.03.074
109. Chen C, Tsow F, Campbell KD, Iglesias R, Forzani E, Tao N (2013) A wireless hybrid chemical sensor for detection of environmental volatile organic compounds. IEEE Sens J 13:1748–1755. https://doi.org/10.1109/JSEN.2013.2239472
110. Caldara M, Colleoni C, Guido E, Re V, Rosace G (2016) Optical monitoring of sweat pH by a textile fabric wearable sensor based on covalently bonded litmus-3-glycidoxypropyltrimethoxysilane coating. Sensors Actuators, B Chem 222:213–220. https://doi.org/10.1016/j.snb.2015.08.073
111. Llorente-Alonso A, Peña-Poza J, De Arcas G, García-Heras M, López JM, Villegas MA (2013) Interface electronic system for measuring air acidity with optical sensors. Sensors Actuators, A Phys 194:67–74. https://doi.org/10.1016/j.sna.2013.01.058
112. Kassal P, Kim J, Kumar R, De Araujo WR, Steinberg IM, Steinberg MD, Wang J (2015) Smart bandage with wireless connectivity for uric acid biosensing as an indicator of wound status. Electrochem Commun 56:6–10. https://doi.org/10.1016/j.elecom.2015.03.018

113. Clément P, Ramos A, Lazaro A, Molina-Luna L, Bittencourt C, Girbau D, Llobet E (2015) Oxygen plasma treated carbon nanotubes for the wireless monitoring of nitrogen dioxide levels. Sensors Actuators, B Chem 208:444–449. https://doi.org/10.1016/j.snb.2014.11.059
114. Warden-Rothman R, Caturegli I, Popik V, Tsourkas A (2013) Sortase-tag expressed protein ligation: combining protein purification and site-specific bioconjugation into a single step. Anal Chem 85:11090–11097. https://doi.org/10.1021/ac402871k
115. Abrar MA, Dong Y, Lee PK, Kim WS (2016) Bendable electro-chemical lactate sensor printed with silver nano-particles. Sci Rep 6:1–9. https://doi.org/10.1038/srep30565
116. Mortellaro M, DeHennis A (2014) Performance characterization of an abiotic and fluorescent-based continuous glucose monitoring system in patients with type 1 diabetes. Biosens Bioelectron 61:227–231. https://doi.org/10.1016/j.bios.2014.05.022
117. Lazaro A, Girbau D, Villarino R (2010) Analysis of vital signs monitoring using an IR-UWB radar. Prog Electromagn Res 100:265–284. https://doi.org/10.2528/PIER09120302
118. Yuechun C, Ganz A (2005) A UWB-based 3D location system for indoor environments. In: 2nd international conference on broadband networks, BROADNETS 2005. 2005, pp 224–232. https://doi.org/10.1109/ICBN.2005.1589737
119. Bandodkar AJ, Jeerapan I, Wang J (2016) Wearable chemical sensors: present challenges and future prospects. ACS Sensors. https://doi.org/10.1021/acssensors.6b00250
120. Xu S, Zhang Y, Cho J, Lee J, Huang X, Jia L, Fan JA, Su Y, Su J, Zhang H, Cheng H, Lu B, Yu C, Chuang C, Kim TI, Song T, Shigeta K, Kang S, Dagdeviren C, Petrov I, Braun PV, Huang Y, Paik U, Rogers JA (2013) Stretchable batteries with self-similar serpentine interconnects and integrated wireless recharging systems. Nat Commun 4:1543–1548. https://doi.org/10.1038/ncomms2553
121. Berchmans S, Bandodkar AJ, Jia W, Ramírez J, Meng YS, Wang J (2014) An epidermal alkaline rechargeable Ag-Zn printable tattoo battery for wearable electronics. J Mater Chem A 2:15788–15795. https://doi.org/10.1039/c4ta03256j
122. Cook-Chennault KA, Thambi N, Sastry AM (2008) Powering MEMS portable devices—a review of non-regenerative and regenerative power supply systems with special emphasis on piezoelectric energy harvesting systems. Smart Mater Struct 17. https://doi.org/10.1088/0964-1726/17/4/043001
123. Kassal P, Steinberg IM, Steinberg MD (2013) Wireless smart tag with potentiometric input for ultra low-power chemical sensing. Sensors Actuators, B Chem 184:254–259. https://doi.org/10.1016/j.snb.2013.04.049
124. De Marcellis A, Depari A, Ferri G, Flammini A, Sisinni E (2013) A CMOS integrated low-voltage low-power time-controlled interface for chemical resistive sensors. Sensors Actuators, B Chem 179:313–318. https://doi.org/10.1016/j.snb.2012.09.104
125. Orecchini G, Yang L, Tentzeris MM, Roselli L (2011) Wearable battery-free active paper-printed RFID tag with human-energy scavenger. 1. https://doi.org/10.1109/mwsym.2011.5973290
126. Zi Y, Lin L, Wang J, Wang S, Chen J, Fan X, Yang PK, Yi F, Wang ZL (2015) Triboelectric-pyroelectric-piezoelectric hybrid cell for high-efficiency energy-harvesting and self-powered sensing. Adv Mater 27:2340–2347. https://doi.org/10.1002/adma.201500121
127. Dong L, Xu C, Li Y, Wu C, Jiang B, Yang Q, Zhou E, Kang F, Yang QH (2016) Simultaneous production of high-performance flexible textile electrodes and fiber electrodes for wearable energy storage. Adv Mater 28:1675–1681. https://doi.org/10.1002/adma.201504747
128. Zhang Y, Zhao Y, Cheng X, Weng W, Ren J, Fang X, Jiang Y, Chen P, Zhang Z, Wang Y, Peng H (2015) Realizing both high energy and high power densities by twisting three carbon-nanotube-based hybrid fibers. Angew Chem Int Ed 54:11177–11182. https://doi.org/10.1002/anie.201506142
129. Kannan R, Garg A (2015) Adaptive sensor fusion technology for mobile and wearable applications. 2015 IEEE Sensors Proceedings, pp 1–4. https://doi.org/10.1109/ICSENS.2015.7370394
130. Sarpeshkar R (2012) Universal principles for ultra low power and energy efficient design. IEEE Trans Circuits Syst II Express Briefs 59:193–198. https://doi.org/10.1109/TCSII.2012.2188451

131. Zhang Y, Zhao Y, Ren J, Weng W, Peng H (2016) Advances in wearable fiber-shaped lithium-ion batteries. Adv Mater 28:4524–4531. https://doi.org/10.1002/adma.201503891
132. Liu L, Yu Y, Yan C, Li K, Zheng Z (2015) Wearable energy-dense and power-dense supercapacitor yarns enabled by scalable graphene-metallic textile composite electrodes. Nat Commun 6:1–9. https://doi.org/10.1038/ncomms8260
133. O'Connor TF, Zaretski AV, Savagatrup S, Printz AD, Wilkes CD, Diaz MI, Sawyer EJ, Lipomi DJ (2016) Wearable organic solar cells with high cyclic bending stability: materials selection criteria. Sol Energy Mater Sol Cells 144:438–444. https://doi.org/10.1016/j.solmat.2015.09.049
134. Kim BJ, Kim DH, Lee YY, Shin HW, Han GS, Hong JS, Mahmood K, Ahn TK, Joo YC, Hong KS, Park NG, Lee S, Jung HS (2015) Highly efficient and bending durable perovskite solar cells: toward a wearable power source. Energy Environ Sci 8:916–921. https://doi.org/10.1039/c4ee02441a
135. Jia W, Valdés-Ramírez G, Bandodkar AJ, Windmiller JR, Wang J (2013) Epidermal biofuel cells: energy harvesting from human perspiration. Angew Chem Int Ed 52:7233–7236. https://doi.org/10.1002/anie.201302922
136. Jia W, Wang X, Imani S, Bandodkar AJ, Ramírez J, Mercier PP, Wang J (2014) Wearable textile biofuel cells for powering electronics. J Mater Chem A 2:18184–18189. https://doi.org/10.1039/c4ta04796f
137. Oh JY, Lee JH, Han SW, Chae SS, Bae EJ, Kang YH, Choi WJ, Cho SY, Lee JO, Baik HK, Lee T (2016) Il: Chemically exfoliated transition metal dichalcogenide nanosheet-based wearable thermoelectric generators. Energy Environ Sci 9:1696–1705. https://doi.org/10.1039/c5ee03813h
138. Lu Z, Zhang H, Mao C, Li CM (2016) Silk fabric-based wearable thermoelectric generator for energy harvesting from the human body. Appl Energy 164:57–63. https://doi.org/10.1016/j.apenergy.2015.11.038
139. Yang JH, Cho HS, Park SH, Song SH, Yun KS, Lee JH (2016) Effect of garment design on piezoelectricity harvesting from joint movement. Smart Mater Struct. https://doi.org/10.1088/0964-1726/25/3/035012
140. Pu X, Li L, Liu M, Jiang C, Du C, Zhao Z, Hu W, Wang ZL (2016) Wearable self-charging power textile based on flexible yarn supercapacitors and fabric nanogenerators. Adv Mater 28:98–105. https://doi.org/10.1002/adma.201504403
141. Seung W, Gupta MK, Lee KY, Shin K-S, Lee J-H, Kim TY, Kim S, Lin J, Kim JH, Kim S-W (2015) Nanopatterned textile-based wearable triboelectric nanogenerator. ACS Nano 9:3501–3509. https://doi.org/10.1021/nn507221f
142. Bandodkar AJ, Jeerapan I, You JM, Nuñez-Flores R, Wang J (2016) Highly stretchable fully-printed CNT-based electrochemical sensors and biofuel cells: combining intrinsic and design-induced stretchability. Nano Lett 16:721–727. https://doi.org/10.1021/acs.nanolett.5b04549
143. Yoon GC, Shin K-S, Gupta MK, Lee KY, Lee J-H, Wang ZL, Kim S-W (2015) High-performance hybrid cell based on an organic photovoltaic device and a direct current piezoelectric nanogenerator. Nano Energy 12:547–555. https://doi.org/10.1016/j.nanoen.2015.01.028
144. Mak WC, Cheung KY, Orban J, Lee CJ, Turner APF, Griffith M (2015) Surface-engineered contact lens as an advanced theranostic platform for modulation and detection of viral infection. ACS Appl Mater Interfaces 7:25487–25494. https://doi.org/10.1021/acsami.5b08644
145. Qiu Y, Li G, Hou Y, Pan Z, Li H, Li W, Liu M, Ye F, Yang X, Zhang Y (2015) Vertically aligned carbon nanotubes on carbon nanofibers: a hierarchical three-dimensional carbon nanostructure for high-energy flexible supercapacitors. Chem Mater 27:1194–1200. https://doi.org/10.1021/cm503784x
146. Wannapob R, Vagin MY, Jeerapan I, Mak WC (2015) Pure nanoscale morphology effect enhancing the energy storage characteristics of processable hierarchical polypyrrole. Langmuir 31:11904–11913. https://doi.org/10.1021/acs.langmuir.5b03318

147. Mesin L (2016) A neural algorithm for the non-uniform and adaptive sampling of biomedical data. Comput Biol Med 71:223–230. https://doi.org/10.1016/j.compbiomed.2016.02.004
148. Wang ZL (2010) Toward self-powered sensor networks. Nano Today 5:512–514. https://doi.org/10.1016/j.nantod.2010.09.001
149. Lou Z, Li L, Wang L, Shen G (2017) Recent progress of self-powered sensing systems for wearable electronics. Small 13:1–27. https://doi.org/10.1002/smll.201701791
150. Shi B, Liu Z, Zheng Q, Meng J, Ouyang H, Zou Y, Jiang D, Qu X, Yu M, Zhao L, Fan Y, Wang ZL, Li Z (2019) Body-integrated self-powered system for wearable and implantable applications. ACS Nano 13:6017–6024. https://doi.org/10.1021/acsnano.9b02233
151. Jeerapan I, Sempionatto JR, Pavinatto A, You JM, Wang J (2016) Stretchable biofuel cells as wearable textile-based self-powered sensors. J Mater Chem A 4:18342–18353. https://doi.org/10.1039/C6TA08358G
152. Yu Y, Zhai J, Xia Y, Dong S (2017) Single wearable sensing energy device based on photoelectric biofuel cells for simultaneous analysis of perspiration and illuminance. Nanoscale 9:11846–11850. https://doi.org/10.1039/c7nr04335j
153. Cho E, Mohammadifar M, Choi S (2017) A self-powered sensor patch for glucose monitoring in sweat. In: Proceedings of IEEE international conference on microelectromechanical systems, pp 366–369. https://doi.org/10.1109/MEMSYS.2017.7863417
154. Garcia SO, Ulyanova YV, Figueroa-Teran R, Bhatt KH, Singhal S, Atanassov P (2016) Wearable sensor system powered by a biofuel cell for detection of lactate levels in sweat. ECS J Solid State Sci Technol 5:M3075–M3081. https://doi.org/10.1149/2.0131608jss
155. Han W, He H, Zhang L, Dong C, Zeng H, Dai Y, Xing L-L, Zhang Y, Xue X (2017) A self-powered wearable non-invasive electronic-skin for perspiration analysis basing on piezo-biosensing unit matrix of enzyme/ZnO nanoarrays. ACS Appl Mater Interfaces. https://doi.org/10.1021/acsami.7b07990
156. He H, Zeng H, Fu Y, Han W, Dai Y, Xing L, Zhang Y, Xue X (2018) A self-powered electronic-skin for real-time perspiration analysis and application in motion state monitoring. J Mater Chem C 6:9624–9630. https://doi.org/10.1039/c8tc03296c
157. Guan H, Zhong T, He H, Zhao T, Xing L, Zhang Y, Xue X (2019) A self-powered wearable sweat-evaporation-biosensing analyzer for building sports big data. Nano Energy 59:754–761. https://doi.org/10.1016/j.nanoen.2019.03.026
158. Reid RC, Mahbub I (2020) Wearable self-powered biosensors. Curr Opin Electrochem 19:55–62. https://doi.org/10.1016/j.coelec.2019.10.002
159. Chen CH, Lee PW, Tsao YH, Lin ZH (2017) Utilization of self-powered electrochemical systems: metallic nanoparticle synthesis and lactate detection. Nano Energy 42:241–248. https://doi.org/10.1016/j.nanoen.2017.10.064
160. Bandodkar AJ, Gutruf P, Choi J, Lee KH, Sekine Y, Reeder JT, Jeang WJ, Aranyosi AJ, Lee SP, Model JB, Ghaffari R, Su CJ, Leshock JP, Ray T, Verrillo A, Thomas K, Krishnamurthi V, Han S, Kim J, Krishnan S, Hang T, Rogers JA (2019) Battery-free, skin-interfaced microfluidic/electronic systems for simultaneous electrochemical, colorimetric, and volumetric analysis of sweat. Sci Adv 5:1–16. https://doi.org/10.1126/sciadv.aav3294
161. Zhang X, Jing Y, Zhai Q, Yu Y, Xing H, Li J, Wang E (2018) Point-of-care diagnoses: flexible patterning technique for self-powered wearable sensors. Anal Chem. https://doi.org/10.1021/acs.analchem.8b02838
162. Aiassa S, Carrara S, Demarchi D (2019) Optimized sampling rate for voltammetry-based electrochemical sensing in wearable and IoT applications. IEEE Sensors Lett 3:1–4. https://doi.org/10.1109/LSENS.2019.2918575
163. Hernandez J, McDuff D, Infante C, Maes P, Quigley K, Picard R (2016) Wearable ESM: differences in the experience sampling method across wearable devices. https://doi.org/10.1145/2935334.2935340
164. Choi J, Ghaffari R, Baker LB, RogersJA (2021) Skin-interfaced systems for sweat collection and analytics. Sci Adv 4:3921. https://doi.org/10.1126/sciadv.aar3921
165. Emaminejad S, Gao W, Wu E, Davies ZA, Yin Yin Nyein H, Challa S, Ryan SP, Fahad HM, Chen K, Shahpar Z, Talebi S, Milla C, Javey A, Davis RW (2017) Autonomous sweat extraction and analysis applied to cystic fibrosis and glucose monitoring using a fully integrated

wearable platform. Proc Natl Acad Sci 114:4625–4630. https://doi.org/10.1073/pnas.1701740114
166. Kim J, Jeerapan I, Imani S, Cho TN, Bandodkar A, Cinti S, Mercier PP, Wang J (2016) Noninvasive alcohol monitoring using a wearable tattoo-based iontophoretic-biosensing system. ACS Sensors 1:1011–1019. https://doi.org/10.1021/acssensors.6b00356
167. García-Carmona L, Martín A, Sempionatto JR, Moreto JR, González MC, Wang J, Escarpa A (2019) Pacifier biosensor: toward noninvasive saliva biomarker monitoring. Anal Chem 91:13883–13891. https://doi.org/10.1021/acs.analchem.9b03379
168. Sempionatto JR, Brazaca LC, García-Carmona L, Bolat G, Campbell AS, Martin A, Tang G, Shah R, Mishra RK, Kim J, Zucolotto V, Escarpa A, Wang J (2019) Eyeglasses-based tear biosensing system: non-invasive detection of alcohol, vitamins and glucose. Biosens Bioelectron 137:161–170. https://doi.org/10.1016/j.bios.2019.04.058
169. Kagie A, Bishop DK, Burdick J, La Belle JT, Dymond R, Felder R, Wang J (2008) Flexible rolled thick-film miniaturized flow-cell for minimally invasive amperometric sensing. Electroanalysis 20:1610–1614. https://doi.org/10.1002/elan.200804253
170. Bandodkar AJ, Jia W, Yardımcı C, Wang X, Ramirez J, Wang J (2015) Tattoo-based noninvasive glucose monitoring: a proof-of-concept study. Anal Chem 87:394–398. https://doi.org/10.1021/ac504300n
171. Liu G-S, Kong Y, Wang Y, Luo Y, Fan X, Xie X, Yang B-R, Wu MX (2020) Microneedles for transdermal diagnostics: Recent advances and new horizons. Biomaterials 232:119740. https://doi.org/10.1016/j.biomaterials.2019.119740
172. Martín A, Kim J, Kurniawan JF, Sempionatto JR, Moreto JR, Tang G, Campbell AS, Shin A, Lee MY, Liu X, Wang J (2017) Epidermal microfluidic electrochemical detection system: enhanced sweat sampling and metabolite detection. ACS Sensors 2:1860–1868. https://doi.org/10.1021/acssensors.7b00729
173. Twine NB, Norton RM, Brothers MC, Hauke A, Gomez EF, Heikenfeld J (2018) Open nanofluidic films with rapid transport and no analyte exchange for ultra-low sample volumes. Lab Chip 18:2816–2825. https://doi.org/10.1039/C8LC00186C
174. Ajami S, Teimouri F (2015) Features and application of wearable biosensors in medical care. J Res Med Sci 20:1208–1215. https://doi.org/10.4103/1735-1995.172991
175. Khanmohammadi A, Jalili Ghazizadeh A, Hashemi P, Afkhami A, Arduini F, Bagheri H (2020) An overview to electrochemical biosensors and sensors for the detection of environmental contaminants. J Iran Chem Soc 17:2429–2447. https://doi.org/10.1007/s13738-020-01940-z
176. McGoldrick LK, Halámek J (2020) Recent advances in noninvasive biosensors for forensics, biometrics, and cybersecurity
177. Nasiri S, Khosravani MR (2020) Progress and challenges in fabrication of wearable sensors for health monitoring. Sensors Actuators, A Phys 312:112105. https://doi.org/10.1016/j.sna.2020.112105
178. Ciui B, Tertis M, Feurdean CN, Ilea A, Sandulescu R, Wang J, Cristea C (2019) Cavitas electrochemical sensor toward detection of N-epsilon (carboxymethyl)lysine in oral cavity. Sensors Actuators B Chem 281:399–407. https://doi.org/10.1016/j.snb.2018.10.096
179. Kim J, Campbell AS, de Ávila BE-F, Wang J (2019) Wearable biosensors for healthcare monitoring. Nat Biotechnol 37:389–406. https://doi.org/10.1038/s41587-019-0045-y
180. Kim J, Campbell AS, Wang J (2018) Wearable non-invasive epidermal glucose sensors: a review. Talanta 177:163–170. https://doi.org/10.1016/j.talanta.2017.08.077
181. Tu J, Torrente-Rodríguez RM, Wang M, Gao W (2020) The era of digital health: a review of portable and wearable affinity biosensors. Adv Funct Mater 30:1906713. https://doi.org/10.1002/adfm.201906713
182. Sharma A, Badea M, Tiwari S, Marty JL (2021) Wearable biosensors: an alternative and practical approach in healthcare and disease monitoring
183. Samson C, Koh A (2020) Stress monitoring and recent advancements in wearable biosensors. https://www.frontiersin.org/article/10.3389/fbioe.2020.01037
184. Sana F, Isselbacher EM, Singh JP, Heist EK, Pathik B, Armoundas AA (2020) Wearable devices for ambulatory cardiac monitoring. J Am Coll Cardiol 75:1582–1592. https://doi.org/10.1016/j.jacc.2020.01.046

185. Ekins S, Mottin M, Ramos PRPS, Sousa BKP, Neves BJ, Foil DH, Zorn KM, Braga RC, Coffee M, Southan C, Puhl AC, Andrade CH (2020) Déjà vu: stimulating open drug discovery for SARS-CoV-2. Drug Discov Today 25:928–941. https://doi.org/10.1016/j.drudis.2020.03.019
186. Cunningham AC, Goh HP, Koh D (2020) Treatment of COVID-19: old tricks for new challenges. Crit Care 24:91. https://doi.org/10.1186/s13054-020-2818-6
187. van Riel D, de Wit E (2020) Next-generation vaccine platforms for COVID-19. Nat Mater 19:810–812. https://doi.org/10.1038/s41563-020-0746-0
188. Gowri A, Ashwin Kumar N, Suresh Anand BS (2021) Recent advances in nanomaterials based biosensors for point of care (PoC) diagnosis of Covid-19—a minireview. TrAC Trends Anal Chem 137:116205. https://doi.org/10.1016/j.trac.2021.116205
189. Un K-C, Wong C-K, Lau Y-M, Lee JC-Y, Tam FC-C, Lai W-H, Lau Y-M, Chen H, Wibowo S, Zhang X, Yan M, Wu E, Chan S-C, Lee S-M, Chow A, Tong RC-F, Majmudar MD, Rajput KS, Hung IF-N, Siu C-W (2021) Observational study on wearable biosensors and machine learning-based remote monitoring of COVID-19 patients. Sci Rep 11:4388. https://doi.org/10.1038/s41598-021-82771-7
190. Ates HC, Yetisen AK, Güder F, Dincer C (2021) Wearable devices for the detection of COVID-19. Nat Electron 4:13–14. https://doi.org/10.1038/s41928-020-00533-1
191. Wang Q, Ling S, Liang X, Wang H, Lu H, Zhang Y (2019) Self-healable multifunctional electronic tattoos based on silk and graphene. Adv Funct Mater 29:1808695. https://doi.org/10.1002/adfm.201808695
192. Kwak MK, Jeong H-E, Suh KY (2011) Rational design and enhanced biocompatibility of a dry adhesive medical skin patch. Adv Mater 23:3949–3953. https://doi.org/10.1002/adma.201101694
193. Magno M, Salvatore GA, Jokic P, Benini L (2019) Self-sustainable smart ring for long-term monitoring of blood oxygenation. IEEE Access 7:115400–115408. https://doi.org/10.1109/ACCESS.2019.2928055
194. Teymourian H, Parrilla M, Sempionatto JR, Montiel NF, Barfidokht A, Van Echelpoel R, De Wael K, Wang J (2020) Wearable electrochemical sensors for the monitoring and screening of drugs. ACS Sensors 5:2679–2700. https://doi.org/10.1021/acssensors.0c01318
195. de Jong M, Sleegers N, Kim J, Van Durme F, Samyn N, Wang J, De Wael K (2016) Electrochemical fingerprint of street samples for fast on-site screening of cocaine in seized drug powders. Chem Sci 7:2364–2370. https://doi.org/10.1039/C5SC04309C
196. Mishra RK, Sempionatto JR, Li Z, Brown C, Galdino NM, Shah R, Liu S, Hubble LJ, Bagot K, Tapert S, Wang J (2020) Simultaneous detection of salivary Δ9-tetrahydrocannabinol and alcohol using a Wearable Electrochemical Ring Sensor. Talanta 211:120757. https://doi.org/10.1016/j.talanta.2020.120757
197. Zhang D, Jiang J, Chen J, Zhang Q, Lu Y, Yao Y, Li S, Logan Liu G, Liu Q (2015) Smartphone-based portable biosensing system using impedance measurement with printed electrodes for 2,4,6-trinitrotoluene (TNT) detection. Biosens Bioelectron 70:81–88. https://doi.org/10.1016/j.bios.2015.03.004

Application of Large-Scale Fabrication Techniques for Development of Electrochemical Biosensors

Giovana Rosso Cagnani and Gisela Ibáñez-Redín

Abstract This chapter overviews the recent advances in the application of large-scale fabrication techniques for developing electrochemical enzymatic biosensors, immunosensors, and nucleic acid-based biosensors. In addition to providing a summary of the main contributions to the field within the last five years, we discussed the existing gaps and challenges that need to be overcome to attain the mass production of electrochemical devices using techniques compatible with roll-to-roll processing. A particular emphasis is placed on the strategies adopted for the fabrication of fully printed enzymatic biosensors.

Keywords Electrochemical biosensors · Large-scale fabrication · Roll-to-roll processing · Printed biosensors · Printing techniques

1 Introduction

There is a growing demand for inexpensive and easy-to-use analytical tests for detecting and quantifying several analytes of environmental, clinical, and industrial relevance. In addition to having excellent analytical performance, such devices must be preferentially disposable, fast, and portable, to be compatible with on-field applications. Electrochemical biosensors have emerged as a promising alternative to meet these requirements, owing to their capability of miniaturization, relatively low instrumentation cost, and the possibility of developing methods with excellent analytical features [80]. In electrochemical biosensors, the signal arising from the interaction between a biological element and a target analyte is transformed into an electrochemical signal proportional to the analyte concentration [81]. Typically, these devices comprise two major components, a biorecognition element (i.e., enzymes,

---

G. R. Cagnani (✉)
São Carlos Institute of Chemistry, University of São Paulo (USP), São Carlos, SP 13560-970, Brazil
e-mail: giovanarosso84@gmail.com

G. Ibáñez-Redín
Sao Carlos Institute of Physics, University of São Paulo (USP), São Carlos, SP 13560-970, Brazil

F. N. Crespilho (ed.), *Advances in Bioelectrochemistry Volume 3*,
https://doi.org/10.1007/978-3-030-97921-8_4

nucleic acids, antibodies, cells, peptides, etc.) and an electrochemical transducer, in combination with an electronic system for signal amplification and processing [34]. An additional advantage of electrochemical biosensors is their versatility, since they can be used with several detection techniques including potentiometry, voltammetry, amperometry, and impedance.

The miniaturization and the production cost of electrochemical sensors have significantly improved over the past decades with the advent of a great variety of large-scale fabrication methods. In particular, screen printing and inkjet printing have been extensively used for producing electrodes for biosensors platforms using different materials and substrates [57, 74]. Devices featuring these kinds of electrodes can be fabricated in different sizes and shapes and designed for ease of automation and integration with technologies such as microfluidic [24]. In addition, the possibility of fabricating electrodes using flexible substrates that can be stretched, deformed, and easily integrated with the skin has driven the research on wearable platforms. Such devices can be used for measuring relevant analytes on biofluids such as sweat, saliva, and tears, offering the possibility of obtaining information in real time on the changes in its chemical composition [69].

The emerging of such interesting applications of electrochemical biosensors has increasingly motivated research into methodologies for their large-scale manufacture, which is crucial for their subsequent commercialization. An example of these methodologies is roll-to-roll processing, which has its origins in the graphic industry but is recently being used in the manufacture of electrochemical devices [7]. Roll-to-roll processing is attractive since it permits depositing solutions and dispersions made of functional materials onto flexible substrates through several printing and coating techniques. Some printing techniques compatible with roll-to-roll processing are rotatory screen printing, inkjet printing, gravure, and flexography. On the other hand, spray and slot-dire coating are examples of coating techniques commonly used in this technology. Besides being useful for electrodes fabrication, roll-to-roll processing can be used for integrating all the stages of biosensors manufacturing, including surfaces functionalization and immobilization of the biorecognition element [95]. Indeed, in the past few years, this approach has been exploited for the high-speed mass fabrication of fully printed sensors and biosensors [13, 82].

This chapter outlines the exciting opportunities offered by different large-scale fabrication techniques for the manufacturing of electrochemical biosensors. Specifically, we will discuss various biosensors platforms on which printing and coating techniques compatible with roll-to-roll processing were used for fabrication. Through this chapter, we will also highlight how the use of methods, such as screen printing, has tremendously influenced the advances in the electrochemical biosensor field, allowing for the emergence and growing of new technologies, including wearable and paper-based biosensors.

2 Enzymatic Biosensors

Electrochemical enzymatic biosensors are among the most widely investigated analytical devices. Indeed, the most commercialized biosensor is the one based on glucose oxidase enzyme, with an amperometric detection, which is used throughout the world to monitor blood glucose levels on diabetes patients [58]. Since the first description of the application of enzymes for sensing in 1962 by Clark and Lyons [20] electrochemical biosensors have been developed for a wide variety of applications, including environmental monitoring, food analysis, and health care [44, 84, 93].

Undoubtedly, the application of large-scale fabrication techniques has had a remarkable influence on the development of biosensors technology. The use of the screen printing for fabricating electrodes and sensors strips for glucose biosensors, in particular, had a huge impact on the commercialization and widespread use of these devices [58]. The interest in screen printing for manufacturing electrochemical devices arose in the early 1980s, but this technology had previously been used in artistic applications and in the production of electronic circuits and solar cells [80]. The main advantage of this technique lies in its simplicity and speed, which allows the mass production of devices with different sizes and shapes using a variety of materials at a low cost. The use of screen printing in electroanalysis has been extensively explored throughout the last decades, and it is well documented in recent literature [4, 9, 74, 80].

One significant achievement in screen-printing technology was the development of inks for low-temperature applications. Nowadays, several commercial inks with low-cure temperatures, based on carbon nanomaterials and metals, are available [58]. Its use has made it possible to replace the traditional ceramics substrates with inexpensive and flexible materials such as paper, plastics, and other biodegradable polymers [22, 73, 88]. This flexibility on the substrate choice has opened the possibility of implementing enzymatic biosensors technology for new and varied applications. The application of electrochemical paper-based analytical devices (ePAD), for example, has become a prominent research area during the last years. These devices owe their success to the low cost, biocompatibility, biodegradability, relative stability, and high abundance of paper [75]. Furthermore, due to its high porosity and capillary action, paper allows the fabrication of platforms in which, along with electrochemical detection, the separation, transport, and preconcentration of the analytes can be performed [8].

Numerous examples of miniaturized and disposable ePADs that use enzymes as biorecognition elements have been described in scientific literature. The pioneering work was reported by Henry's group in 2009 [28], which fabricated a paper-based microfluidic device for multiplexed detection of glucose, lactate, and uric acid. More recently, Cinti and coworkers developed a paper-based screen-printed electrochemical biosensor to quantify paraoxon [18]. The device combines a SPE with a microfluidic cell that is fabricated by wax printing. The butyrylcholinesterase enzyme and the substrate, butyrylthiocholine, were incorporated into the system by depositing them on nitrocellulose and paper membranes. This approach allows the reagent-free detection of paraoxon with a limit of detection of 3 μg/L. Another interesting application involving butyrylcholinesterase was proposed by Caratelli and coworkers [14]; they developed an origami paper-based electrochemical device for monitoring the activity of this enzyme on blood in the presence of drugs used for Alzheimer's disease. Figure 1a shows the device, which combines a paper-based SPE modified with carbon black/Prussian blue nanocomposite, a membrane impregnated with butyrylthiocholine, and a Vivid™ plasma separation GX membrane. The sensor was capable of monitoring the inhibitory effect of physostigmine, rivastigmine, and donepezil and showed great potential to be applied in customized drug administration.

The sensitive and rapid detection of sulfur mustard agent, one of the most dangerous chemical weapons, was also achieved using origami paper-based electrochemical biosensors (Fig. 1b) [21]. Paper-based SPEs were prepared using graphite inks modified with a carbon black/Prussian blue nanocomposite, and then, the choline oxidase enzyme and substrate (choline chloride) were preloaded by depositing them onto separate areas of the origami. The detection of sulfur mustard was done by monitoring the enzyme inhibition using amperometry. This sensor exhibited limits of detection of 1 mM and 0.019 g·min m^{-3} for liquid and aerosol phases, respectively.

The emergence of techniques compatible with large-scale manufacturing has also prompted research in wearable sensors. These devices can be integrated within the human body and used to continuously monitor several physiological parameters such as hearth rate, blood pressure, temperature, and biomarkers concentration [51]. Wearable enzymatic biosensors have been fabricated using different flexible substrates including polyimide, polyethylene terephthalate (PET), paper, polyurethane, textile, and transfer tattoos, among others [21, 29, 39, 52, 53, 70, 86]. The potential of these devices has been investigated for a variety of applications ranging from athlete monitoring to personalized medicine [31, 70, 71, 91, 92]. Screen-printing technique has been applied for the fabrication of wearable electrochemical ring biosensor [54]. The device, whose fabrication process is shown in Fig. 1c, contains a voltametric sensor for detecting tetrahydrocannabinol and an amperometric biosensor for monitoring alcohol levels. The sensor allows the simultaneous detection of tetrahydrocannabinol and alcohol in saliva samples within 3 min, without significant interference from the saliva components.

An interesting approach in wearable biosensors is the incorporation of iontophoretic interfaces for sweat stimulation. This is particularly relevant for health monitoring and medical diagnosis, where sweat stimulation through traditional approaches (i.e., increment of the environmental temperature, exercise, etc.) is not always possible [51]. Epidermal devices with integrated iontophoresis electrodes and electrochemical sensors have been fabricated through screen printing. These devices have been applied for monitoring several different types of analytes in sweat and interstitial fluid, including glucose, lactate, caffeine, and alcohol. An interesting application of this technology was recently proposed by Sempionatto and coworkers [68]. They developed an epidermal biosensor for monitoring the dynamics of sweat vitamin C after the intake of supplements or fruit juices. The device, fabricated on a flexible polyurethane substrate, combines a screen-printed biosensor with two

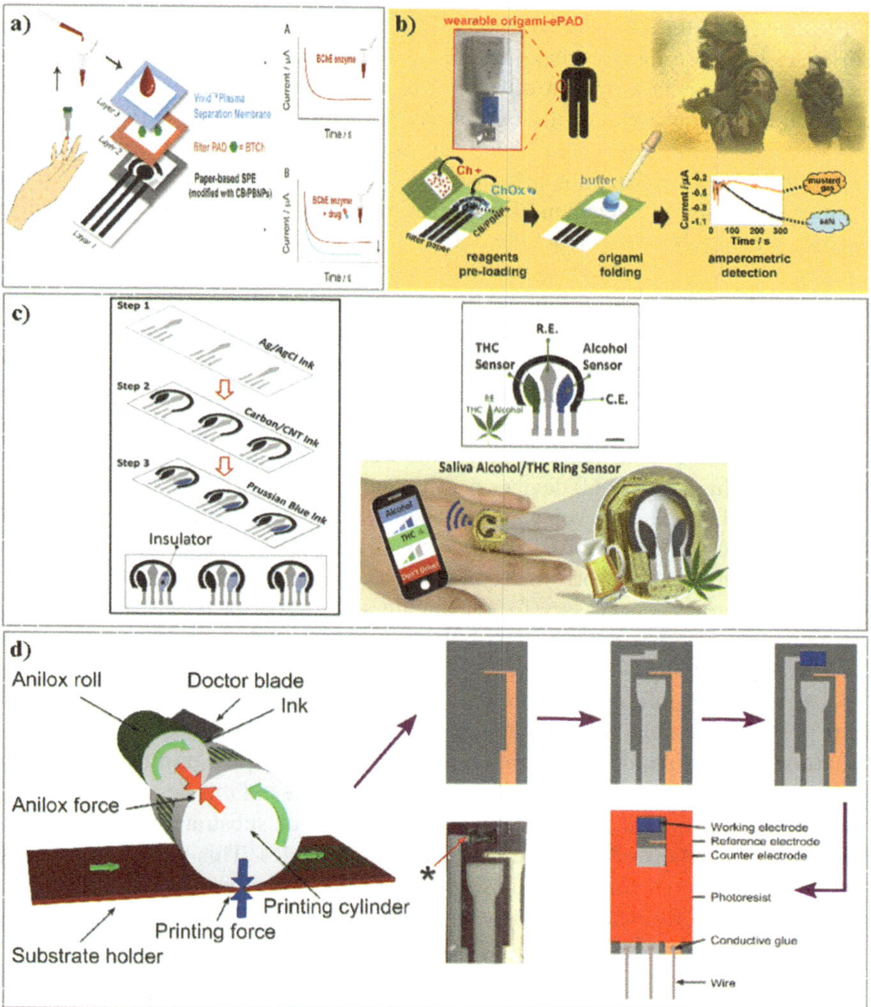

Fig. 1 **a** Origami paper-based electrochemical biosensor for detection of cholinesterase inhibitors. Reprinted with permission from [14]. Copyright 2020, Elsevier. **b** Wearable origami paper-based biosensor for detection of sulfur mustard. Reprinted with permission from [21]. Copyright 2019, Elsevier. **c** Schematic illustration of the fabrication process, main components, and the application of the electrochemical ring sensor for simultaneous detection of Δ9-tetrahydrocannabinol and alcohol. Reprinted (adapted) with permission from [54]. Copyright 2020, Elsevier. **d** Schematic representation of the flexographic printer and steps for the fabrication of electrodes modified with ZnO nanowires. Reprinted (adapted) with permission from [30]. Copyright 2017, Elsevier

silver electrodes for sweat stimulation, using pilocarpine-based iontophoresis. The biosensor based on the ascorbate oxidase enzyme was successfully used for monitoring the sweat concentration of vitamin C on different subjects, showing great potential for noninvasive nutrition status assessments.

Conventional screen printing has been the most widely used fabrication technique to create electrodes for biosensors development. However, in recent years, the interest in fabrication techniques compatible with roll-to-roll processing such as rotatory screen printing, gravure, flexography, and slot-die coating has significantly increased [7, 30, 62] [16, 59, 82]. This trend is motivated by the need to develop automatic, fast, and high-throughput methodologies that allow integrating different stages of production of the devices. Mass fabrication of microfluidic sensing patch for monitoring Na^+, glucose, and sweat rate has been recently achieved using roll-to-roll rotatory screen printing [59]. These devices combine a printed three-electrode system for glucose and Na^+ sensing with a spiral-shaped impedimetric system for sweat rate measurement. These components were printed together on a 100-m web length with a high rate of 60 devices per minute.

Besides being able to develop electrodes patterning, large-scale fabrication techniques are also suitable for modifying electrodes surfaces. This is a critical step on the fabrication of biosensors since, in most cases, the incorporation of polymers and nanomaterials is necessary for enhancing sensing performance and providing functional groups for enzyme immobilization [1]. However, most currently used modification techniques (*i.e.*, drop casting, layer-by-layer, electrodeposition, etc.) involve time-consuming and laborious procedures, being thus incompatible with large-scale fabrication. Benson and coworkers proposed a methodology for direct patterning of gold nanoparticles using roll-to-roll flexographic printing with application in glucose biosensing [10]. The fabrication process of this device involves the flexographic printing of a carbon layer onto a polyimide substrate, followed by the printing of a customized ink based on gold nanoparticles. This platform was used for immobilization of glucose oxidase enzyme using cysteamine and glutaraldehyde. The biosensor showed good performance for detecting glucose, with a sensitivity of 5.52 μAmM^{-1} $cm^{-2}$and a limit of detection of 26 μM.

An interesting application of roll-to-roll flexographic printing for glucose biosensing was proposed by Fung and coworkers [30]. They used commercially available Ag/AgCl and carbon inks to pattern reference, counter and carbon electrodes onto a polyimide substrate. After curing, a layer of zinc acetate precursor ink was printed onto the working electrode to produce ZnO seeds (Fig. 1d). This layer allowed the selective growth of ZnO nanowires using a hydrothermal method. This platform was functionalized with glucose oxidase enzyme and employed as a biosensor for the detection of glucose. The fabricated device exhibited a sensitivity of 1.2 μAmM^{-1} cm^{-2} with a limit of detection of 46 μM.

Another tendency in electrochemical biosensors is the development of fully printed devices, where the biomolecules are also immobilized using printing or coating techniques. This approach requires the formulation of specific bioinks which, in addition to having the appropriate properties for printing (i.e., surface tension and viscosity), must provide an adequate environment to maintain the structure of the biomolecules [61]. Depending on the biomolecule and the detection method, this is not always an easy task, which explains why some techniques remain relatively unexplored for these types of applications. Additives commonly used in bioinks formulations are surfactants and polymeric viscous modifiers (e.g., polyethyleneglycol, carboxymethyl cellulose, and chitosan). Non-ionic surfactants such as Triton X-100 and those from the Tween family are preferred, since most proteins are charged and their structure can be affected in the presence of charged molecules [61].

From the earliest use of screen printing in electroanalysis, a variety of fully printed enzymatic biosensors have been fabricated. Enzymes such as acetyl cholinesterase, lactate oxidase, and glucose oxidase have been screen printed onto different substrates [32, 33, 39, 50, 68]. For this, several approaches have been proposed, including mixing the enzymes with commercially available carbon inks and preparing specific inks based on polymers and sol–gel materials [3]. Despite being suitable for enzyme printing, the screen-printing technique suffers from several limitations such as the large amount of ink required. Furthermore, the deposition of different layers of materials demands using specific masks for each layer and complicated alignment steps. Inkjet printing has proven to be a more suitable technique allowing for depositing individual drops only onto specific spots without using any mask [48, 77]. This is particularly valuable for fabricating multiplexed devices, which require the immobilization of different types of biomolecules. Li and coworkers developed a process to fabricate multiplexed biosensors for detecting glucose, lactate, and triglycerides using inkjet printing [46]. The fabrication process illustrated in Fig. 2a consisted on printing conductive polyaniline hydrogels and enzymes on the top of screen-printed electrodes. Using this approach, 96 electrodes can be modified within 5 min.

Kang and coworkers fabricated all-inkjet-printed glucose biosensors using inks based on single-walled carbon nanotubes (SWCNTs) and M13 phage [40]. M13 phage was used at this study due to its capability to form a complex containing carbon nanomaterials with high dispersibility and stability in aqueous medium. The custom ink was used to pattern a three-electrode system onto glossy paper. The working electrode was then coated with a layer of positively charged polyethyleneimine followed by the printing of ten layers of a bioink containing glucose oxidase and phosphate buffer (Fig. 2b). The biosensor allows for the detection of glucose with high stability and a limit of detection of 20 μM. Furthermore, the device was used for monitoring glucose concentration in human sweat, showing an excellent correlation with commercially available assay kits.

Bihar and coworkers also developed a totally inkjet-printed electrochemical glucose biosensor [11]. Figure 2c shows a picture of the devices which are formed by three electrodes deposited onto paper substrate using PEDOT:PSS ink. After electrodes patterning, an aqueous ink containing glucose oxidase, chitosan, N-(3-dimethylaminopropyl)-N′-ethylcarbodiimide hydrochloride (EDC), sulfo-N hydroxysuccinimide (sulfo-NHS) and redox mediator ferrocene, was printed on the top of the working electrode. Finally, a thin layer of Nafion was printed over the sensor area. This disposable and biodegradable device exhibits good sensitivity to glucose detection on saliva within a concentration range from 0.025 to 0.9 mM.

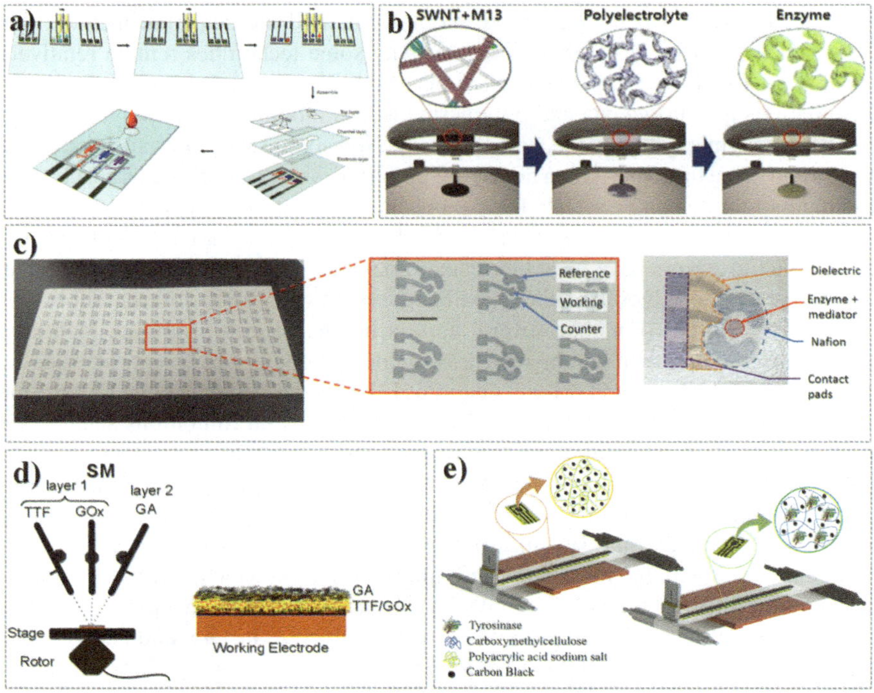

Fig. 2 Fully printed electrochemical enzymatic biosensors. **a** Schematic representation of the fabrication of inkjet-printed multiplexed biosensors for detection of glucose, lactate, and triglycerides based on conductive hydrogels. Reprinted (adapted) with permission from [46]. Copyright 2018, American Chemical Society. **b** Steps for the fabrication of all-inkjet-printed glucose biosensor using inks based on SWCNTs and M13 phage. Reprinted (adapted) with permission from [40]. Copyright 2020, American Chemical Society. **c** Photograph of glucose biosensors inkjet-printed on paper using PEDOT:PSS inks. Reprinted (adapted) with permission from [11]. Copyright 2018, The Authors. **d** Schematic diagram of the spray mixing deposition of tetrathiafulvalene, glucose oxidase, and glutaraldehyde onto carbon working electrodes. Reprinted (adapted) with permission from [66]. Copyright 2018, Elsevier. **e** Illustration of roll-to-roll slot-die coating of SPCEs with carbon black-based inks and bioinks. Reprinted with permission from [13]. Copyright 2020, Elsevier

Although in a smaller portion, enzymatic biosensors based on spray coating have also been developed [23, 66]. Samavat and coworkers proposed a technique for deposition of sensing layers containing enzyme and redox mediators using spray coating [66]. The methodology is illustrated in Fig. 2d and consists on simultaneously depositing an ethanolic solution of tetrathiafulvalene (TTF) and an aqueous solution of glucose oxidase (GOx) onto a polyimide modified with carbon ink. These surfaces were coated with a layer of glutaraldehyde to improve enzyme adherence. The biosensor allows for the sensitive detection of glucose within a concentration range from 2 to 6 mM.

The use of flexographic and gravure printing for enzyme deposition in the development of electrochemical biosensors is much more limited. This is due to the high shear forces and pressure to which inks are subjected in these techniques. These harsh conditions can significantly affect the enzyme structure, leading to its denaturation and loss of activity [61]. Therefore, the formulation of enzymatic inks for gravure and flexography printing is particularly challenging. Stable enzymatic inks for flexographic printing have been developed [35, 67]. However, to our knowledge, they have not yet been used in electrochemical biosensors. Pavinatto and coworkers developed a flexible and fully printed biosensor using gravure to deposit the enzymatic active layer [62]. The enzymatic inks containing phosphate buffer, Tyrosinase, Triton X-100, trehalose, and carboxymethylcellulose, were printed using gravure on the top of interdigitated Au electrodes. Although after printing, the enzyme preserved only 15% of its initial activity, the device was successfully applied on the impedimetric detection of pyrogallol with a limit of detection of 200 μM.

An alternative technique for depositing biomolecules without significantly affecting their structure is slot-die coating. Recently, Cagnani and coworkers described the fabrication of printed sensors and biosensors for detecting dopamine and catechol using roll-to-roll slot-die deposition [13]. The enzymatic biosensor was fabricated by coating inks based on tyrosinase and carbon black onto the surfaces of SPEs. Since slot-die technique is not suitable for applications that require specific patterning, the deposition area was delimited using adhesive templates (Fig. 2e). Negligible changes in the enzymatic activity measured by UV–Vis were observed, which is due to the absence of high temperatures or mechanical stress during the deposition process. The biosensor exhibited excellent performance for detecting catechol, with high stability and a low limit of detection of 1.9 μM.

3 Immunosensors

Antibody-based sensors, also called immunosensors, detect changes generated by the specific antibody–antigen interaction and transform them into an analytical signal. In these devices, the detection can be done using different strategies that can or not involve the use of labels. Electrochemical sandwich-type immunosensors, which are based on the sandwich enzyme-linked immunosorbent assay (ELISA), have been extensively used since their emergence at the end of the 1990s [2]. In these biosensors, primary antibodies are immobilized at the working electrode surface and used to capture the antigen (target analyte), which selectively interacts with a secondary antibody. Magnetic nanoparticles can also be used as a surface for the immobilization of primary antibodies, and this works as a strategy for removing interferences that may impair an accurate analysis [49]. The secondary antibody is often attached to a detection label responsible for generating an electrochemical signal [24]. Commonly, enzymes, nanoparticles, and conductive polymers are used as labels to achieve ultrasensitive detection [89]. The electrochemical techniques used for detection are similar to those employed in enzymatic biosensors, such as amperometry, differential pulse voltammetry (DPV), square wave voltammetry (SWV), and cyclic voltammetry.

One of the main advantages of electrochemical immunosensors is the possibility of performing label-free detection. In label-free electrochemical immunosensors, the antigen–antibody interaction is measured without using secondary antibodies, which results in a faster and simpler alternative, with better cost effect. Electrochemical techniques commonly employed in label-free immunosensors are as follows: DPV, SWV, and electrochemical impedance spectroscopy (EIS). The latter technique in particular is widely used since it allows to monitor interaction events at the electrode surface without using redox probes [79].

Electrochemical immunosensors have been used to detect a large number of analytes, including disease biomarkers, drugs, environmental pollutants, and pathogens [26, 37, 63, 83, 90]. The advent of a variety of manufacturing techniques and the possibility of using paper and polymeric substrates has significantly improved the production costs and the capability of miniaturization of these devices [24]. Nowadays, new research pathways have been opened, and the potential of electrochemical immunosensors is being explored in the development of miniaturized, flexible, and disposable devices, suitable for wearable and point-of-care applications [89]. SPE fabricated using polymeric substrates has been extensively used for applications in immunosensors. Cheng and coworkers developed battery-free, wireless, and flexible electrochemical patches for detection of sweat cortisol using screen-printed carbon electrodes deposited onto a PET substrate (Fig. 3a) [17]. Gold nanoparticles were electrodeposited onto the working electrode surface, followed by the adsorption of a layer of thiol polyethylene glycol carboxyl. The carboxylic groups on the polymer film were used for the covalent immobilization of anti-cortisol antibodies. This label-free immunosensor was used to measure cortisol using a DPV technique. The device exhibited a limit of detection of 7.47 nM, which is within the concentration range of cortisol in the human sweat. Additionally, the electrochemical patch was used for in situ detection of cortisol on volunteers' sweat, showing an excellent correlation with the ELISA results.

Paper-based immunosensors, fabricated by screen printing, have also been reported. Prasad and coworkers developed a paper-based electrochemical immunosensor for pseudopodium-enriched atypical kinase one (PEAK1), a biomarker for pancreatic ductal adenocarcinoma [64]. The sensitive immunosensor was fabricated by printing a three-electrode system onto chromatographic paper pre-treated with a SU-8 photoresist (Fig. 3b). The working electrode was coated with graphene oxide (GO) followed by the covalent immobilization of anti-PEAK1 antibodies. A sandwich immunoassay was conducted for detecting PEAK1 using Au nanoparticles modified with anti-PEAK1 antibodies as a label. The immunosensor exhibited a wide linear range from 10 to 106 pg mL^{-1} and a low limit of detection of 10 pg mL^{-1}.

Electrochemical immunosensors fabricated by inkjet printing were reported to have low-cost and point-of care detection of several compounds and biomolecules. These devices have been manufactured using commercially available and custom-fabricated inks based on metallic nanoparticles, conducting polymers, and carbon

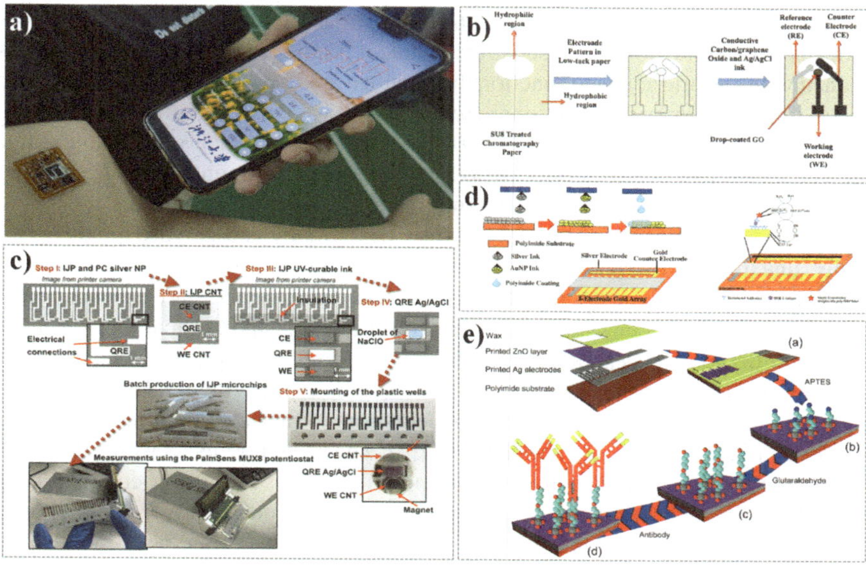

Fig. 3 Electrochemical immunosensors with electrodes patterned using large-scale fabrication techniques. **a** Photograph of the wireless flexible electrochemical patch for in situ analysis of sweat cortisol. Reprinted (adapted) with permission from [17]. Copyright 2021, Elsevier. **b** Schematic representation of the fabrication of disposable paper-based electrodes for application in PEAK1 detection. Reprinted with permission from [64]. Copyright 2019, Elsevier. **c** Illustrations and photographs describing the fabrication of inkjet-printed electrochemical microchips for point-of-care immunoassays. Reprinted with permission from [38]. Copyright 2017, Elsevier. **d** Schematic representation of the fabrication and detection mechanism of the fully inkjet-printed electrochemical array platform for detecting HER-2 biomarker. Reprinted (adapted) with permission from [15]. Copyright 2018, Elsevier. **e** Schematic representation of the fabrication of immunosensors for detecting human cytomegalovirus antigen. Reprinted with permission from [5]. Copyright 2016, American Chemical Society

nanomaterials [36, 60, 94, 96]. Jović and coworkers reported the large-scale fabrication of electrochemical microchips for point-of-care immunoassays using inkjet printing [38]. Figure 3c shows a schematic representation of the fabrication process, which involves the sequential deposition of silver, CNT, and insulator inks onto PET substrate. Each plate contains eight independent three-electrode cells fabricated with carbon nanotubes and Ag/AgCl-based inks. After printing, the reaction area of each device was delimited by adhering high-density polyethylene sheets with well-defined holes. The fabricated devices were applied for detecting biomarker thyroid-stimulating hormone in urine and pollutant atrazine in water samples using magnetic immunoassays.

Inkjet-printed electrodes based on metallic inks are an excellent platform for immobilizing antibodies in immunosensors applications. Carvajal and coworkers developed an inkjet-printed immunosensor for detection of the breast cancer biomarker HER-2 [15]. Figure 3d shows a scheme of the platform which consists of an eight-electrode gold array, a counter electrode, and a silver reference electrode deposited onto polyimide substrate by inkjet printing. The electrodes array was cleaned with sulfuric acid and functionalized with a monolayer of mercaptopropionic acid to introduce carboxyl functional groups, which were used for covalent immobilization of anti-HER-2 antibodies. The array was used in combination with a microfluidic channel for the detection of HER-2 in a sandwich immunoassay. The device exhibited an excellent reproducibility and low limit of detection of 12 pg mL^{-1}.

Unlike screen printing and inkjet printing, flexography and gravure remain practically unexplored for patterning electrodes for applications in electrochemical immunosensors. This is mainly due to the relatively higher start-up and prototyping cost of these techniques, which requires the design and fabrication of specific printing rollers [95]. Despite this, a few papers from literature report the use of immunosensors with electrodes printed by flexography and gravure. For instance, a laboratory-scale flexographic printer has been used to fabricate immunosensors for detecting human cytomegalovirus (HCMV) [5]. The immunosensors were fabricated by printing interdigitated electrodes onto polyimide substrate using commercially available silver inks. After curing, a water-based zinc acetate ink was printed onto the surface and thermally annealed at 300 °C for 30 min. This process allows the production of nanostructured ZnO thin films. Specific antibodies for recognizing HCM pp65 were immobilized onto the electrode surface. Pp65 antigen was selected as a target because of its high abundance within extracellular virus particles of HCMV. Figure 3e shows a schematic representation of the fabrication process of the immunosensor, which exhibited high selectivity for pp65 with a limit of detection of 5 pg mL^{-1}. A rotogravure technique was used to fabricate interdigitated electrodes for biosensing at the Western Michigan University's pilot plant [65]. The electrodes with finger dimensions of 200 μm width and spacing were printed onto a PET substrate using conductive silver nanoparticle-based ink. The device was tested for detecting different concentrations of D-proline and mouse IgG using impedance spectroscopy.

Spray coating and inkjet printing have been used to modify electrodes surfaces with nanomaterials suitable for immobilizing antibodies. A flexible immunosensor for the detection of histamine was developed using SPEs coated with SWCNTs [72]. Silver electrodes were screen printed onto a PET substrate and an ink containing surfactant sodium dodecyl sulfate, SWCNTs, and water was spray deposited onto the working electrode. Anti-histamine antibodies were covalently immobilized at the electrode surface, and histamine was detected in a competitive immunoassay using secondary antibodies labeled with horseradish peroxidase (HRP). The devices exhibited a high stability and a wide linear detection range from 0.005 to 50 ng mL^{-1}. Boonkaew and coworkers developed a sensitive immunosensor for detecting ferritin biomarker using paper-based SPEs [12]. Inkjet printing was used to modify the working electrodes with graphene oxide inks, prepared using N-methyl-2-pyrolidone as solvent. The monoclonal antibodies specific to ferritin were covalently immobilized on the working electrode. This electrochemical device exhibited good selectivity and a limit of detection of 0.19 ng mL^{-1}, by DPV.

Sensitive and selective detection of the HT-2 mycotoxin was achieved using immunosensors with inkjet-printed electrochemically reduced graphene oxide microelectrodes [43]. The counter and reference electrodes, as well as the conductive tracks, were screen printed onto a poly(ethylene 2,6-naphthalate) substrate using commercially available carbon and Ag/AgCl inks. A water-based graphene oxide ink was formulated and used to print a working electrode with dimensions of 78 μm in width and 30 nm in height using inkjet printing. The printed graphene oxide electrode was electrochemically reduced and used as a platform for immobilizing anti-HT-2 antibodies. This biosensor exhibited a low limit of detection of 1.6 ng mL^{-1} and a dynamic range of 6.3 – 100 ng mL^{-1}.

4 Biosensors Using Nucleic Acids as Biorecognition Element

Aptamers, single-stranded DNA (ssDNA), and RNA sequences are also used as biorecognition elements in electrochemical biosensors [89]. DNA-based biosensors (Genosensors) are suitable for performing fast, sensitive, and inexpensive identification of mutations using low sample volumes. This is particularly relevant for applications in clinical analysis, since DNA mutations are related to several diseases and its detection may assist in early diagnosis. In these platforms, ssDNA sequences are immobilized at the electrode surface, and the hybridization with its complementary sequence (cDNA or target) is detected. As in the case of immunosensors, the target can be detected using labeled or label-free methodologies. Electrochemical genosensors have been produced for screening several diseases including cancer, Alzheimer, and others neurodegenerative disorders [6, 41, 56]. Different fabrication techniques and substrates have been used seeking the production of easy-to-use and inexpensive devices, which allows working with low sample volumes. An important contribution in this regard was made by Cinti and coworkers, who developed paper-based strips for detecting single and double-stranded DNA (Fig. 4a) [19]. They used filter paper and copy paper as a substrate for fabricating SPEs, which were functionalized with gold nanoparticles and DNA probes. The authors performed an extensive characterization of the electrochemical and analytical performance of both platforms, without observing a significant difference among them. This is particularly important, since, as reported by the authors, the use of office paper can provide a 30% savings over the device production costs.

Besides being able to monitor mutations in human DNA, genosensors can also be useful for detecting genetic material of pathogens including viruses and bacteria. An interesting biosensor for detecting HPV was developed by Teengam and coworkers, which use peptide nucleic acids (PNAs) molecules as bioreceptor [78]. PNAs are synthetic nucleic acids analogs in which the sugar-phosphate backbone is replaced by repeating N-(2-aminoethyl)-glycine aminoethyl)-glycine units [45].

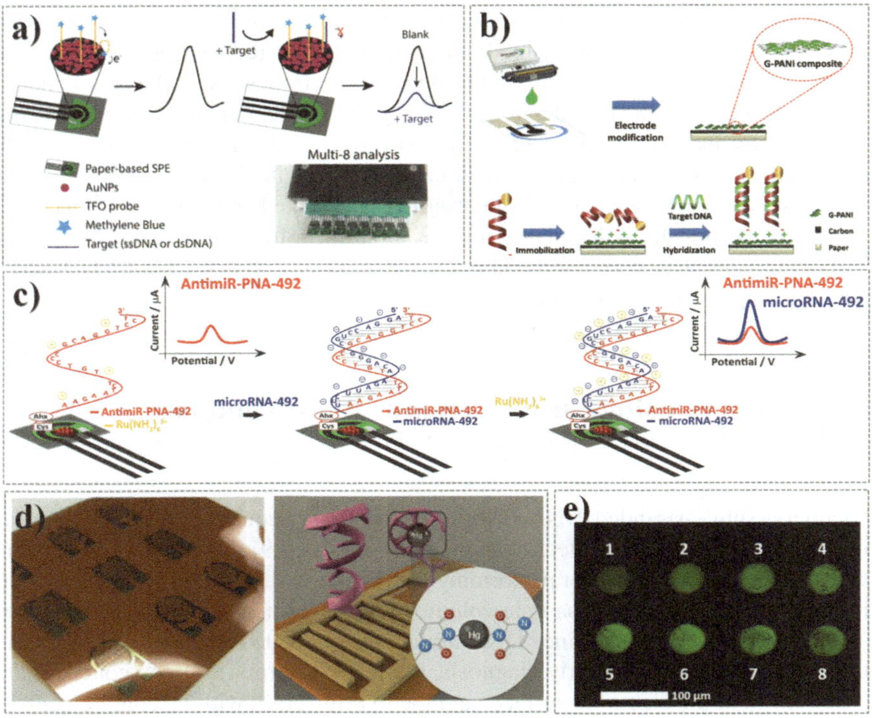

Fig. 4 Electrochemical biosensors using nuclei acids and nucleic acids analogs as bioreceptors. **a** Schematic representation of the paper-based DNA platform fabricated by screen printing. Reprinted with permission from [19]. Copyright 2018, American Chemical Society. **b** Schematic illustration of the electrode modification and detection mechanism of the paper-based electrochemical sensor for detecting human papillomavirus. Reprinted (Adapted) with permission from [78]. Copyright 2016, Elsevier. **c** Schematic representation of the detection of miRNA-492 using paper-based electrochemical PNA biosensor. Reprinted (Adapted) with permission from [55]. Copyright 2020, Elsevier. **d** Photograph of inkjet-printed interdigitated electrodes patterned on Kapton substrate and illustration of the mechanism for detection of Hg^{2+}. Reprinted (Adapted) with permission from [25]. Copyright 2019, Elsevier. **e** Fluorescence image of substrates modified with different printed layers (from 1 to 8) of carbon nanotubes–aptamer bioinks. Reprinted (Adapted) with permission from [42]. Copyright 2020, The Authors

These molecules have attracted significant attention for the development of biosensors due to their chemical stability and higher resistance to nucleases compared with ssDNA. The proposed biosensor was fabricated using electrodes modified by inkjet printing of graphene–polyaniline inks. The anthraquinone-labeled pyrrolidinyl PNA probe was immobilized onto the modified electrode surface through electrostatic attraction, and HPV 16 sequences were detected using SWV. Figure 4b depicts a schematic representation of the detection mechanism. The hybridization impairs the interaction between anthraquinone and the electrode surface, thus leading to the signal decrease. Under optimized conditions, the obtained limit of detection for HPV 16 was 2.3 nM, within a linear range of 10–200 nM.

Disposable and easy-to-use PNAs-based biosensors have also been developed for detecting miRNA-492, a biomarker for pancreatic ductal adenocarcinoma [55]. A three-electrode system was screen printed onto office paper previously impermeabilized with wax. The working electrode was coated with gold nanoparticles and functionalized with thiolated PNAs molecules specifically designed to recognize miRNA-492. In this platform, miRNA-492 was detected through a "signal on" approach by monitoring the adsorption of $Ru(NH_3)_6^{3+}$ with DPV measurements (Fig. 4c). Under optimized conditions, the biosensor exhibited a high selectivity and a low limit of detection of 19 nM.

Aptamers can adopt three-dimensional structures, which allows them to bind different targets analytes with high affinity [89]. These short oligonucleotides are synthetized through the selection evolution of ligands by exponential enrichment (SELEX) technique, in which receptors to target analytes of different sizes can be produced without using animals or cell cultures [47]. One advantage of aptamers is its relatively high stability when compared to antibodies or enzymes, which makes aptasensors more amenable for analysis under harsh conditions. Disposable and easy-to-use electrochemical aptasensors have been fabricated using different materials and applied for detecting a wide range of analytes including bacteria, organic pollutants, cancer cells, exosomes, and heavy metals [27, 76, 85, 87]. For instance, Diaz-Amaya and coworkers reported the impedimetric detection of Hg^{2+} in water and organic solvents using aptasensors based on inkjet-printed electrodes (Fig. 4d) [25]. Interdigitated gold electrodes were printed onto Kapton® films using commercially available inks. An aptameric DNA sequence modified with thiol group was designed for specific recognizing Hg^{2+} and immobilized at the electrode surface by drop casting. This device exhibited an excellent performance toward the detection of Hg^{2+} in water and dimethyl sulfoxide, with a limit of detection of 0.005 ppm, which is below the minimal sensitivity required for practical applications.

The relatively high stability of aptamers under harsh conditions has allowed the fabrication of fully printed aptasensors using nanomaterials containing inks. Khan and coworkers fabricated an inkjet-printed aptasensor for the selective detection of lysozyme [42] inks containing MWCNTs, lysozyme binding aptamer, and N-methyl pyrrolidone, and they were prepared and deposited onto the surface of carbon electrodes. Ink patternability was first investigated by fluorescence imaging using aptamers labeled with a fluorophore. Printed patterns with different numbers of layers are presented in Fig. 4e, where the correlation among the number of layers and the intensity is clearly seen. The devices with five printed layers of bioink were used for the impedimetric detection of lysozyme, with a limit of detection of 90 ng mL^{-1}.

5 Conclusion and Outlook

In this chapter, we provided an overview of the advances in the application of large-scale fabrication methods for the development of electrochemical biosensors. From the survey in the literature, we found that screen printing and inkjet printing have been more used for the research stage since stencils are easy to fabricate and desktop printers can be adapted to deposit custom inks. Furthermore, throughout the years, the efforts have focused primarily on improving the device's efficacy (i.e., long-term stability, non-specific binding, limit of detection, and sensitivity). Nowadays, many devices have been well tested, and there is a growing interest in methodologies to translate the technology into commercial applications, which is likely to explain the increasing interest in roll-to-roll processing for electroanalysis. Significant progress within the last years on ink formulations has allowed the high-throughput fabrication of electrochemical devices using rotatory screen printing, flexography, and gravure. However, translating electrochemical biosensors into market requires not only methodologies for patterning electrodes, but also automated procedures that incorporate stages of nanomaterials deposition, functionalization, and immobilization of biomolecules. Thus, a portion of current research is being directed toward the development of methodologies compatible with roll-to-roll processing for direct deposition of functional materials and biomolecules onto the surfaces of the printed electrodes. Considering the recent advances in the field [82], the ideal process for mass manufacturing of electrochemical systems will not be based on a single technique, but on the integration of different mastered methods.

Fully printed biosensors have gained significant attention during the last years, as evidenced by the increasing number of publications in this area. Inkjet has been preferred for printing biomolecules, but other techniques, including gravure, slot-die, and spray coating, have also proven to be suitable for such purposes. Establishing methodologies for large-scale printing of biomolecules, however, requires more research on functionalization strategies and bioinks formulation. Even though a few of the papers reviewed in this chapter have addressed topics such as ink rheology and biomolecules activity after printing, an extensive characterization of inks properties such as long-term stability still has to be done. This lack of information on aspects extremely relevant for obtaining devices with optimal performance offers excellent opportunities for future research in this area.

Overall, new progress in the application of large-scale deposition techniques for manufacturing electrochemical biosensors offers great hope that these devices will soon surpass the frontier of academic research and be in widespread use for different highly specialized applications including disease diagnosis, personalized nutrition, pathogens detection, food safety, and work security.

Acknowledgements Giovana Rosso Cagnani gratefully acknowledges the postdoctoral grant by MeDiCo Network CAPES-Brazil (23038.003012/2020-16), and Gisela Ibáñez-Redín acknowledges the technical training grant by FAPESP (2020/14906-5).

References

1. Ahmad R et al (2018) Deposition of nanomaterials: a crucial step in biosensor fabrication. Mater Today Commun 17:289–321. https://doi.org/10.1016/j.mtcomm.2018.09.024
2. Aizawa M et al (1976) An enzyme immunosensor for IgG. J Solid-Phase Biochem 1(4):319–328. https://doi.org/10.1007/BF02990970
3. Albareda-Sirvent M, Merkoçi A, Alegret S (2000) Configurations used in the design of screen-printed enzymatic biosensors. A review. Sens Actuators, B Chem 69(1):153–163. https://doi.org/10.1016/S0925-4005(00)00536-0
4. Arduini F et al (2016) Electrochemical biosensors based on nanomodified screen-printed electrodes: recent applications in clinical analysis. Trends Anal Chem 79:114–126. https://doi.org/10.1016/j.trac.2016.01.032
5. Assaifan AK et al (2016) SI nanotextured surface on flexographic printed ZnO thin films for low-cost non-faradaic biosensors. ACS Appl Mater Interfaces 8(49):33802–33810. https://doi.org/10.1021/acsami.6b11640
6. Azimzadeh M, Nasirizadeh N (2017) Early detection of Alzheimer ' s disease using a biosensor based on electrochemically-reduced graphene oxide and gold nanowires for the quanti fi cation of serum microRNA-137. RSC Adv 7:55709–55719. https://doi.org/10.1039/c7ra09767k
7. Bariya M et al (2018) Roll-to-Roll gravure printed electrochemical sensors for wearable and medical devices. ACS Nano 12(7):6978–6987. https://doi.org/10.1021/acsnano.8b02505
8. Batista Deroco P et al (2020) Paper-based electrochemical sensing devices. In" Comprehensive analytical chemistry, 1st edn. Elsevier B.V., pp 91–137. https://doi.org/10.1016/bs.coac.2019.11.001
9. Beitollahi H et al (2020) Applications of electrochemical sensors and biosensors based on modified screen-printed electrodes: A review. Anal Methods 12(12):1547–1560. https://doi.org/10.1039/c9ay02598g
10. Benson J et al (2015) Direct patterning of gold nanoparticles using flexographic printing for biosensing applications. Nanoscale Res Lett 10(1):1–8. https://doi.org/10.1186/s11671-015-0835-1
11. Bihar E et al (2018) A fully inkjet-printed disposable glucose sensor on paper. NPJ Flex Electron 2(1):1–8. https://doi.org/10.1038/s41528-018-0044-y
12. Boonkaew S et al (2020) Cost-effective paper-based electrochemical immunosensor using a label-free assay for sensitive detection of ferritin. Analyst 145(14):5019–5026. https://doi.org/10.1039/d0an00564a
13. Cagnani GR et al (2020) Fully-printed electrochemical sensors made with flexible screen-printed electrodes modified by roll-to-roll slot-die coating. Biosens Bioelectron 165:112428. https://doi.org/10.1016/j.bios.2020.112428
14. Caratelli V et al (2020) Precision medicine in Alzheimer's disease: An origami paper-based electrochemical device for cholinesterase inhibitors. Biosens Bioelectron 165(June):112411. https://doi.org/10.1016/j.bios.2020.112411
15. Carvajal S et al (2018) Disposable inkjet-printed electrochemical platform for detection of clinically relevant HER-2 breast cancer biomarker. Biosens Bioelectron 104:158–162. https://doi.org/10.1016/j.bios.2018.01.003
16. Chekol F et al (2019) Roll-to-Roll printed PEDOT/PSS/GO plastic film for electrochemical determination of carbofuran. Electroanalysis 31(6):1104–1111. https://doi.org/10.1002/elan.201800883
17. Cheng C et al (2021) Battery-free, wireless, and flexible electrochemical patch for in situ analysis of sweat cortisol via near field communication. Biosens Bioelectron 172:112782. https://doi.org/10.1016/j.bios.2020.112782
18. Cinti S et al (2017) Fully integrated ready-to-use paper-based electrochemical biosensor to detect nerve agents. Biosens Bioelectron 93:46–51. https://doi.org/10.1016/j.bios.2016.10.091
19. Cinti S et al (2018) Paper-based strips for the electrochemical detection of single and double stranded DNA. Anal Chem 90(22):13680–13686. https://doi.org/10.1021/acs.analchem.8b04052

20. Clark LC, Lyons C (1962) Electrode systems for continuous monitoring in cardiovascular surgery. Ann N Y Acad Sci 102:29–45
21. Colozza N et al (2019) A wearable origami-like paper-based electrochemical biosensor for sulfur mustard detection. Biosens Bioelectron 129(January):15–23. https://doi.org/10.1016/j.bios.2019.01.002
22. Colozza N, Caratelli V, Moscone D (2021) Origami paper-based electrochemical (bio) sensors: state of the art and perspective. Biosensors 11(328):1–29
23. Cortina M et al (2006) Urea impedimetric biosensor based on polymer degradation onto interdigitated electrodes. Sens Actuators, B Chem 118(1–2):84–89. https://doi.org/10.1016/j.snb.2006.04.062
24. Dhanapala L et al (2020) Printed electrodes in microfluidic arrays for cancer biomarker protein detection. Biosensors 10(9):1–27. https://doi.org/10.3390/bios10090115
25. Diaz-Amaya S et al (2019) Inkjet printed electrochemical aptasensor for detection of Hg2+ in organic solvents. Electrochim Acta 316:33–42. https://doi.org/10.1016/j.electacta.2019.05.079
26. Dixit CK et al (2015) Electrochemistry-based approaches to low cost, high sensitivity, automated, multiplexed protein immunoassays for cancer diagnostics. The Analyst 141(2):536–547. https://doi.org/10.1039/c5an01829c
27. Dou B et al (2019) Aptamer-functionalized and gold nanoparticle array-decorated magnetic graphene nanosheets enable multiplexed and sensitive electrochemical detection of rare circulating tumor cells in whole blood. Anal Chem 91(16):10792–10799. https://doi.org/10.1021/acs.analchem.9b02403
28. Dungchai W, Chailapakul O, Henry CS (2009) Electrochemical detection for paper-based microfluidics. Anal Chem 81(14):5821–5826. https://doi.org/10.1021/ac9007573
29. Emaminejad S et al (2017) Autonomous sweat extraction and analysis applied to cystic fibrosis and glucose monitoring using a fully integrated wearable platform. Proc Natl Acad Sci USA 114(18):4625–4630. https://doi.org/10.1073/pnas.1701740114
30. Fung CM et al (2017) Facile fabrication of electrochemical ZnO nanowire glucose biosensor using roll to roll printing technique. Sens Actuators, B Chem 247:807–813. https://doi.org/10.1016/j.snb.2017.03.105
31. García-Carmona L et al (2019) Pacifier biosensor: toward noninvasive saliva biomarker monitoring. Anal Chem 91(21):13883–13891. https://doi.org/10.1021/acs.analchem.9b03379
32. Hart AL, Collier WA (1998) Stability and function of screen printed electrodes, based on cholinesterase, stabilised by a co-polymer/sugar alcohol mixture. Sens Actuators, B Chem 53(1–2):111–115. https://doi.org/10.1016/S0925-4005(98)00300-1
33. Hart AL, Turner APF, Hopcroft D (1996) On the use of screen- and ink-jet printing to produce amperometric enzyme electrodes for lactate. Biosens Bioelectron 11(3):263–270. https://doi.org/10.1016/0956-5663(96)88413-2
34. Hoyos-Nogués M, Gil FJ, Mas-Moruno C (2018) Antimicrobial peptides: Powerful biorecognition elements to detect bacteria in biosensing technologies. Molecules 23(7):1–24. https://doi.org/10.3390/molecules23071683
35. Hurley BRA et al (2013) Flexographic printing of trametes versicolor laccase for indicator applications. Packag Technol Sci 29(January):399–412. https://doi.org/10.1002/pts
36. Jarošová R et al (2019) Inkjet-printed carbon nanotube electrodes for measuring pyocyanin and uric acid in a wound fluid simulant and culture media. Anal Chem 91(14):8835–8844. https://doi.org/10.1021/acs.analchem.8b05591
37. Jones AL et al (2021) Prostate cancer diagnosis in the clinic using an 8-protein biomarker panel. Anal Chem 93(2):1059–1067. https://doi.org/10.1021/acs.analchem.0c04034
38. Jović M et al (2017) Inkjet-printed microtiter plates for portable electrochemical immunoassays. J Electroanal Chem 786:69–76. https://doi.org/10.1016/j.jelechem.2016.12.051
39. Kagie A et al (2008) Flexible rolled thick-film miniaturized flow-cell for minimally invasive amperometric sensing. Electroanalysis 20(14):1610–1614. https://doi.org/10.1002/elan.200804253
40. Kang TH et al (2020) All-inkjet-printed flexible nanobio-devices with efficient electrochemical coupling using amphiphilic biomaterials. ACS Appl Mater Interfaces 12(21):24231–24241. https://doi.org/10.1021/acsami.0c02596

41. Khalilzadeh B et al (2019) Development of a reliable microRNA based electrochemical genosensor for monitoring of miR-146a, as key regulatory agent of neurodegenerative disease. Int J Biol Macromol 134:695–703. https://doi.org/10.1016/j.ijbiomac.2019.05.061
42. Khan NI, Maddaus AG, Song E (2018) A low-cost inkjet-printed aptamer-based electrochemical biosensor for the selective detection of lysozyme. Biosensors 8(7):2–18. https://doi.org/10.3390/bios8010007
43. Kudr J et al (2020) Inkjet-printed electrochemically reduced graphene oxide microelectrode as a platform for HT-2 mycotoxin immunoenzymatic biosensing. Biosens Bioelectron 156(February):112109. https://doi.org/10.1016/j.bios.2020.112109
44. Kurbanoglu S, Erkmen C, Uslu B (2020) Frontiers in electrochemical enzyme based biosensors for food and drug analysis. TrAC-Trends Anal Chem 124:115809. https://doi.org/10.1016/j.trac.2020.115809
45. Lai Q et al (2021) Application strategies of peptide nucleic acids toward electrochemical nucleic acid sensors. Analyst 146:5822–5835. https://doi.org/10.1039/d1an00765c
46. Li L et al (2018) All inkjet-printed amperometric multiplexed biosensors based on nanostructured conductive hydrogel electrodes. Nano Lett 18(6):3322–3327. https://doi.org/10.1021/acs.nanolett.8b00003
47. Liu M et al (2019) Aptasensors for pesticide detection. Biosens Bioelectron 130:174–184. https://doi.org/10.1016/j.bios.2019.01.006
48. Mass M et al (2021) Fully inkjet-printed biosensors fabricated with a highly stable ink based on carbon nanotubes and enzyme-functionalized nanoparticles. Nanomaterials 11(7). https://doi.org/10.3390/nano11071645
49. Materón EM et al (2021) Magnetic nanoparticles in biomedical applications: a review. Appl Surface Sci Adv 6:100163. https://doi.org/10.1016/j.apsadv.2021.100163
50. Mersal GAM, Khodari M, Bilitewski U (2004) Optimisation of the composition of a screen-printed acrylate polymer enzyme layer with respect to an improved selectivity and stability of enzyme electrodes. Biosens Bioelectron 20:305–314. https://doi.org/10.1016/j.bios.2004.01.023
51. Min J et al (2021) Wearable electrochemical biosensors in North America. Biosens Bioelectron 172:112750. https://doi.org/10.1016/j.bios.2020.112750
52. Mishra RK et al (2017) Wearable flexible and stretchable glove biosensor for on-site detection of organophosphorus chemical threats. ACS Sensors 2(4):553–561. https://doi.org/10.1021/acssensors.7b00051
53. Mishra RK et al (2018) Wearable potentiometric tattoo biosensor for on-body detection of G-type nerve agents simulants. Sens Actuators, B Chem 273(May):966–972. https://doi.org/10.1016/j.snb.2018.07.001
54. Mishra RK et al (2020) Simultaneous detection of salivary Δ9-tetrahydrocannabinol and alcohol using a wearable electrochemical ring sensor. Talanta 211:120757. https://doi.org/10.1016/j.talanta.2020.120757
55. Moccia M et al (2020) Paper-based electrochemical peptide nucleic acid (PNA) biosensor for detection of miRNA-492: a pancreatic ductal adenocarcinoma biomarker. Biosens Bioelectron 165(June):112371. https://doi.org/10.1016/j.bios.2020.112371
56. Mohammadi H, Yammouri G, Amine A (2019) Current advances in electrochemical genosensors for detecting microRNA cancer markers. Curr Opin Electrochem 16:96–105. https://doi.org/10.1016/j.coelec.2019.04.030
57. Moya A et al (2017) Inkjet-printed electrochemical sensors. Curr Opin Electrochem 3(1):29–39. https://doi.org/10.1016/j.coelec.2017.05.003
58. Newman JD, Turner APF (2005) Home blood glucose biosensors: a commercial perspective. Biosens Bioelectron 20(12):2435–2453. https://doi.org/10.1016/j.bios.2004.11.012
59. Nyein HYY et al (2019) Regional and correlative sweat analysis using high-throughput microfluidic sensing patches toward decoding sweat. Sci Adv 5(8):1–12. https://doi.org/10.1126/sciadv.aaw9906
60. Otieno BA et al (2016) Cancer diagnostics via ultrasensitive multiplexed detection of parathyroid hormone-related peptides with a microfluidic immunoarray. Anal Chem 88(18):9269–9275. https://doi.org/10.1021/acs.analchem.6b02637

61. Pavinatto FJ (2015) Biosensors with printed active layer. Display and Imaging 2(1):69–91
62. Pavinatto FJ, Paschoal CWA, Arias AC (2015) Printed and flexible biosensor for antioxidants using interdigitated ink-jetted electrodes and gravure-deposited active layer. Biosens Bioelectron 67:553–559. https://doi.org/10.1016/j.bios.2014.09.039
63. Pollap A, Kochana J (2019) Electrochemical immunosensors for antibiotic detection. Biosensors 9(2). https://doi.org/10.3390/bios9020061
64. Prasad KS et al (2020) A low-cost nanomaterial-based electrochemical immunosensor on paper for high-sensitivity early detection of pancreatic cancer. Sens Actuators, B Chem 305:127516. https://doi.org/10.1016/j.snb.2019.127516
65. Reddy ASG et al (2011) Gravure printed electrochemical biosensor. Procedia Eng 25:956–959. https://doi.org/10.1016/j.proeng.2011.12.235
66. Samavat S et al (2018) Uniform sensing layer of immiscible enzyme-mediator compounds developed via a spray aerosol mixing technique towards low cost minimally invasive microneedle continuous glucose monitoring devices. Biosens Bioelectron 118(July):224–230. https://doi.org/10.1016/j.bios.2018.07.054
67. Savolainen A et al (2011) Printing of polymer microcapsules for enzyme immobilization on paper substrate. Biomacromol 12(6):2008–2015. https://doi.org/10.1021/bm2003434
68. Sempionatto JR, Brazaca LC et al (2019) Eyeglasses-based tear biosensing system: non-invasive detection of alcohol, vitamins and glucose. Biosens Bioelectron 137(March):161–170. https://doi.org/10.1016/j.bios.2019.04.058
69. Sempionatto JR, Jeerapan I et al (2019) Wearable chemical sensors: emerging systems for on-body analytical chemistry. Anal Chem. https://doi.org/10.1021/acs.analchem.9b04668
70. Sempionatto JR et al (2020) Epidermal enzymatic biosensors for sweat vitamin C: toward personalized nutrition. ACS Sensors 5(6):1804–1813. https://doi.org/10.1021/acssensors.0c00604
71. Sempionatto JR et al (2021) An epidermal patch for the simultaneous monitoring of haemodynamic and metabolic biomarkers. Nat Biomed Eng 5:737–748. https://doi.org/10.1038/s41551-021-00685-1
72. Shkodra B et al (2020) Flexible and printed electrochemical immunosensor coated with oxygen plasma treated SWCNTs for histamine detection. Biosensors 10(4):1–12. https://doi.org/10.3390/bios10040035
73. Silva RR et al (2020) Microbial nanocellulose adherent to human skin used in electrochemical sensors to detect metal ions and biomarkers in sweat. Talanta 218:121153. https://doi.org/10.1016/j.talanta.2020.121153
74. Smart A et al (2020) Screen-printed carbon based biosensors and their applications in agri-food safety. TrAC—Trends Anal Chem 127:115898. https://doi.org/10.1016/j.trac.2020.115898
75. Solhi E, Hasanzadeh M, Babaie P (2020) Electrochemical paper-based analytical devices (ePADs) toward biosensing: recent advances and challenges in bioanalysis. Anal Methods 12(11):1398–1414. https://doi.org/10.1039/d0ay00117a
76. Subjakova V et al (2021) Advances in electrochemical aptasensors and immunosensors for detection of bacterial pathogens in food. Electrochim Acta 389:138724. https://doi.org/10.1016/j.electacta.2021.138724
77. Suman EO et al (2011) Chronocoulometric determination of urea in human serum using an inkjet printed biosensor. Anal Chim Acta 697(1–2):98–102. https://doi.org/10.1016/j.aca.2011.04.036
78. Teengam P et al (2017) Electrochemical paper-based peptide nucleic acid biosensor for detecting human papillomavirus. Anal Chim Acta 952:32–40. https://doi.org/10.1016/j.aca.2016.11.071
79. Thivina V et al (2016) Design and fabrication of Interdigitated Electrode (IDE) for detection of Ganoderma boninense. In: IEEE international conference on semiconductor electronics, proceedings, ICSE, 2016-Septe, pp 50–53. https://doi.org/10.1109/SMELEC.2016.7573588
80. Trojanowicz M (2016) Impact of nanotechnology on design of advanced screen-printed electrodes for different analytical applications. TrAC-Trends Anal Chem 84:22–47. https://doi.org/10.1016/j.trac.2016.03.027

81. Turner APF (2013) Biosensors: sense and sensibility. Chem Soc Rev 42(8):3184. https://doi.org/10.1039/c3cs35528d
82. Ulloa AM et al (2021) Roll-to-Roll manufactured sensors for nitroaromatic organophosphorus pesticides detection. ACS Appl Mater Interfaces 13(30):35961–35971. https://doi.org/10.1021/acsami.1c08700
83. Valverde A et al (2018) Determination of cadherin-17 in tumor tissues of different metastatic grade using a single incubation-step amperometric immunosensor. Anal Chem 90(18):11161–11167. https://doi.org/10.1021/acs.analchem.8b03506
84. Verma ML, Rani V (2021) Biosensors for toxic metals, polychlorinated biphenyls, biological oxygen demand, endocrine disruptors, hormones, dioxin, phenolic and organophosphorus compounds: a review. Environ Chem Lett 19(2):1657–1666. https://doi.org/10.1007/s10311-020-01116-4
85. Wang L et al (2020) Recent advances in the development of electrochemical aptasensors for detection of heavy metals in food. Biosens Bioelectron 147:111777. https://doi.org/10.1016/j.bios.2019.111777
86. Wang R et al (2021) Stretchable gold fiber-based wearable textile electrochemical biosensor for lactate monitoring in sweat. Talanta 222:121484. https://doi.org/10.1016/j.talanta.2020.121484
87. Wang S et al (2017) Aptasensor with expanded nucleotide using DNA nanotetrahedra for electrochemical detection of cancerous exosomes. ACS Nano 11(4):3943–3949. https://doi.org/10.1021/acsnano.7b00373
88. Wang SQ et al (2016) Flexible substrate-based devices for point-of-care diagnostics. Trends Biotechnol 34(11):909–921. https://doi.org/10.1016/j.tibtech.2016.05.009
89. Wehmeyer KR et al (2021) Electrochemical affinity assays/sensors: brief history and current status. Annu Rev Anal Chem 14:109–131. https://doi.org/10.1146/annurev-anchem-061417-125655
90. Wilson D et al (2019) Electrical detection of pathogenic bacteria in food samples using information visualization methods with a sensor based on magnetic nanoparticles functionalized with antimicrobial peptides. Talanta 194:611–618. https://doi.org/10.1016/j.talanta.2018.10.089
91. Wiorek A et al (2020) Epidermal patch with glucose biosensor: pH and temperature correction toward more accurate sweat analysis during sport practice. Anal Chem 92(14):10153–10161. https://doi.org/10.1021/acs.analchem.0c02211
92. Xuan X et al (2021) Lactate biosensing for reliable on-body sweat analysis. ACS Sensors 6(7):2763–2771. https://doi.org/10.1021/acssensors.1c01009
93. Yang A, Yan F (2021) Flexible electrochemical biosensors for health monitoring. ACS Appl Electron Mater 3(1):53–67. https://doi.org/10.1021/acsaelm.0c00534
94. Yang B et al (2018) Sensitive detection of thyroid stimulating hormone by inkjet printed microchip with a double signal amplification strategy. Chin Chem Lett 29(12):1879–1882. https://doi.org/10.1016/j.cclet.2018.01.042
95. Zavanelli N, Kim J, Yeo WH (2021) Recent advances in high-throughput nanomaterial manufacturing for hybrid flexible bioelectronics. Materials 14(11). https://doi.org/10.3390/ma14112973
96. Zea M et al (2019) All inkjet printing sensor device on paper: for immunosensors applications. IEEE Xplore, pp 2472–2475. https://doi.org/10.1109/TRANSDUCERS.2019.8808473

Screen-Printed Electrochemical Sensors and Biosensors for Detection of Biomarkers

Ava Gevaerd, Luiz R. G. Silva, Tiago Almeida Silva, Luiz H. Marcolino-Junior, Márcio F. Bergamini, and Bruno Campos Janegitz

Abstract In this chapter, the use of miniaturized 2D-printed electrochemical devices for the detection of biomarkers is explored. This section reports the different types and forms of construction of 2D electrochemical devices, as well as their architecture, employed materials and production methods. In addition, different models are explored, such as electrodes obtained by screening, made from stamps and wearable sensors, including the modification with anchoring of different species, such as thin films, metallic nanoparticles, DNA strands, antibodies, enzymes, among others. Furthermore, the application of the devices and their advantages will be discussed in detail, as well as the future prospects for the use of these devices.

Keywords 2D printing · Electrochemistry · Biomarkers · Screen-printed electrodes · Biosensors

1 Introduction

The monitoring of biomarkers in clinical practice as well in laboratory researches has been widely employed over the last years due to its presence and/or concentration level be accepted almost without question as a primary endpoint in several clinical trials. With the integration of different systems, it is possible to create a variety of devices that bring together several functions for an analytical procedure. Also, with

A. Gevaerd · L. H. Marcolino-Junior · M. F. Bergamini
Laboratório de Sensores Eletroquímicos (LabSensE), Departamento de Química, Universidade Federal Do Paraná (UFPR), Curitiba, PR 81531-980, Brazil

A. Gevaerd
Hilab, Rua José Altair Possebom, 800, Curitiba, PR 81270-185, Brazil

T. A. Silva
Department of Chemistry, Federal University of Viçosa, Viçosa, MG 36570-900, Brazil

L. R. G. Silva · B. C. Janegitz (✉)
Laboratory of Sensors, Nanomedicine, and Nanostructured Materials, Federal University of São Carlos, Araras, SP 13600-970, Brazil
e-mail: brunocj@ufscar.br

F. N. Crespilho (ed.), *Advances in Bioelectrochemistry Volume 3*,
https://doi.org/10.1007/978-3-030-97921-8_5

the advancement of micro and nanotechnologies, it is possible to build platforms that come closer and closer to the dimensions of the biological world, facilitating the study and determination of some biological systems, processes, and species. This fact explains the great worldwide interest in the study of the so-called miniaturized electroanalytical devices systems produced based on 2D platforms, which encourage this chapter.

1.1 Biomarkers

Biomarkers refer to measurable biological parameters and quantifiable that can serve as indicators for related assessments to health and patient physiology. A target biomolecule can be detected in the blood or other body fluid that is associated with a sign of a normal or abnormal process, or of a condition or disease can be used as a biomarker. The use of biomarkers for clinical analysis purposes is easier and cheaper than directly measuring the "final disease," with the advantage of the possibility of measurement and determination in a shorter period since these marker species usually reveal themselves in advance.

They can be used in the tracking, diagnosis, characterization, and monitoring of diseases; as prognostic indicators and for therapeutic interventions; to predict and treat adverse drug reactions, as well provide further information on the disease or intervention into account, among other possibilities (Fig. 1). Good biomarkers should be measured with little or no variability, should be a signal ratio, and can meet several criteria for different uses or exhibit specific characteristics that allow its particular use [1].

Pletcher et al. [2] describe that there are three fundamental mechanisms, in the context of clinical care, based on the determination, measurement, and quantification of a biomarker that can improve an individual's health. Figure 2 shows a representative scheme of the mechanisms described by the authors, which are as follows: help the patient to understand their illness or disease risk and thus directly improve their quality of life and/or mental health; motivate the patient to make behavioral changes that improve health; help a doctor to make a better clinical decision about the use of a treatment that leads to an improvement in the patient's health. The goal of determining biomarkers is to develop screening tests that can detect diseases before the evolution of symptoms or at a stage where they are most effectively treated since altered levels in specific types of biomarkers can be closely related to several illnesses [3, 4].

Biological fluids (e.g., blood, saliva, and urine) are the most used samples in laboratory tests and provide useful biological information for the monitoring of various diseases. They are composed of a wide range of biomolecules, such as enzymes, hormones, carbohydrates, nucleic acids, proteins, lipids, salts, and various other organic and inorganic compounds that can be used as biomarkers that reflect and indicate an individual health status.

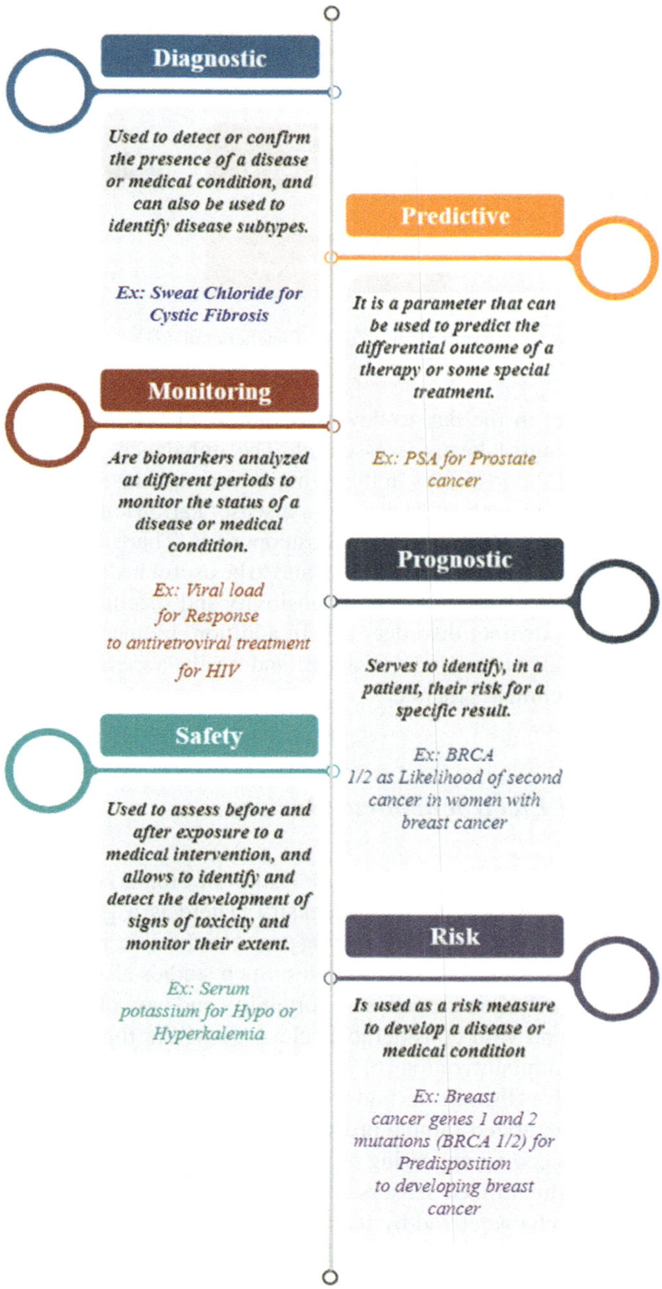

Fig. 1 Types of biomarkers. Adapted from [1]

Fig. 2 Goals for biomarker determination based on Pletcher et al. [2]

Each class is vital in the day-to-day activities of each organism and therefore continuous monitoring of levels is essential. The imbalance, whether due to the absence or excess of these species in the living organism, causes a series of biological disorders to appear, such as Alzheimer's disease, Parkinson's disease, diabetes, heart attack, pregnancy complications, and osteoporosis. There are hundreds of cataloged biomarkers described in the literature, and to be useful in diagnosis, biomarkers need to provide sufficiently high levels of sensitivity and specificity in detecting and correctly classifying distinct disorders [5]. In addition, biomarkers must be reproducible, reliable, inexpensive, non-invasive, and easily accessible to ensure their application in daily clinical practice.

1.2 2D-Printed Electrochemical Devices

The search for different platforms for the detection of biomarkers has been a great motivation for research in analytical chemistry. There is a growing demand for devices able to offer continuous and fast response and stable measurements using small volumes and low concentration samples. Such studies also point to the development of systems with characteristics of portability and ease of use and are closely linked to another trend with considerable relevance within the scope of analytical chemistry, which is miniaturization [6, 7].

The concept involves the sensors and biosensors miniaturization since they can be also coupled to miniaturized sample preprocessing and detection systems. Electrochemical techniques end up appearing as a promising and highly recommended set of transducers for miniaturized sensors and biosensors. In general, electrochemical transduction can be characterized by its rapid response, high analytical sensitivity, wide linearity, repeatability, and adaptability to miniaturized and portable analysis systems. The availability of commercial electrochemical instrumentation adapted for fieldwork also emphasizes the role of electrochemical techniques in the search for more robust and efficient miniaturized devices.

For the manufacture of electrochemical detection devices that meet the above requirements and that make use of low-cost materials, the technique based on 2D printing has been increasingly explored. Through the application of this technology, it is possible to design complete electrochemical measurement systems, in an easy and scalable way, and that present functional characteristics of operation and design according to the demand to be used. Thus, the most relevant technical aspects about the fabrication and evaluation of electrochemical devices manufactured by 2D printing are discussed below.

The purpose of 2D printing is directed to the manufacture of screen-printed electrodes, among the different possibilities, screen printing (SP) is an appropriate strategy for the preparation of conductive films on an inert support (glass, ceramic, or polymer) [8]. The use of this strategy involves simple and inexpensive equipment and processes, yet providing a highly reproducible manufacturing method, and the SP has proven highly compatible for the development of disposable sensors since the 1990s. A classic configuration of SPEs is shown in Fig. 3. In this case, the SPE has typical dimensions of $3 \times 1 \times 0.05$ cm and consists of three electrodes: working electrode (WE), counter electrode (CE), and pseudo-reference electrode (PRE). Despite this typical configuration, available from various manufacturers, laboratory protocols can be implemented to obtain other designs. Within this context, sensors/detectors

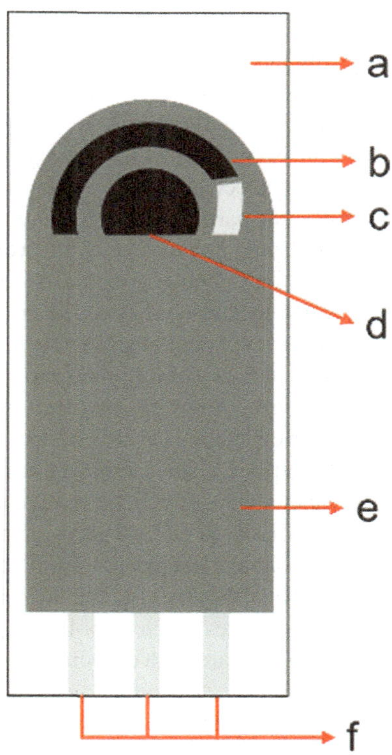

Fig. 3 Representation of a typical SPE configuration: **a** inert solid substrate; **b** counter electrode; **c** pseudo-reference electrode; **d** working electrode; **e** insulator layer; **f** conducting paths for connection. Adapted from [11]

employed for analytical purposes stand out as one of the lines of research that can most evolve with the use of these technologies [9, 10].

The main 2D-printing methods employed to manufacture electrodes and electrochemical devices include inkjet printing, screen printing, and roll-to-roll (R2R) printing. The basic working principle of each of these techniques can be seen in the schemes of Fig. 4. Inkjet printing is a droplet ejection-based technique, in which droplets are typically generated either by piezoelectric actuation (piezoelectric inkjet printing) or thermally (thermal inkjet printing) [12]. The inkjet printers can be operated in drop-on-demand or continuous modes, as distinguished in Fig. 4a. In the drop-on-demand inkjet printing mode, the target substrate is maintained on a fixed position while a moveable print head ejects droplets at the desired positions [12]. On the other hand, in the continuous mode, the fixed and moving parts are inverted, i.e., a fixed print head provides a constant stream of droplets on the moveable target substrate being, therefore, a preferred operation mode for industrial applications [12]. In the case of screen printing, it is carried out by using a stencil screen (made of a mesh of silk, fabric, synthetic fibers, or metal threads) [13]. Thus, the moving of a roller or squeegee across the screen ensures the passage of the ink through the threads of the woven mesh and reaching the inert solid support of the print. Figure 4b illustrates the screen-printing process. To increase productivity, the R2D printing method outlined in Fig. 4c can be adapted. Another approach for obtaining 2D-printed electrodes was reported by Windmiller et al. [14] using the pattern transfer technique. Figure 5 details the steps involved in the stamp transfer fabrication of electrodes. By using this simple and low-cost technique, sensors and biosensors can be directly prepared on different types of irregular surfaces and with different morphologies.

In preparing the SPEs, a conductive ink with appropriate viscosity (of 3.0–10 Pa s) is forced to flow through a screen, which defines the shape and size of the desired electrodes [11]. Different stencil screens are normally prepared to print the different parts of the SPE, and the printing is performed sequentially after the thermal drying and curing of the previous layer. Thus, the manufacturing process of SPEs involves choosing the desired design for the printing screen and the materials (ink and substrate) and studying any subsequent modifications of the surface of the electrodes via chemical and/or electrochemical treatment, or deposition of thin films of nanomaterials or biological recognition species.

Considering the steps involved in the preparation of SPEs, numerous researches have been developed to study new compositions of conductive inks, as well as printing solid substrates, always aiming at high analytical performance associated with the use of cheaper and less toxic alternative materials. As for the substrate, great prominence can be offered to SPEs based on paper and polymers. Both give rise to flexible and disposable electrodes, whose adhesion (especially in the case of paper) of the conductive ink is very efficient. Moreover, paper is biodegradable, and polymers from recycling sources (such as polyethylene terephthalate (PET) from soda bottles) can be used, which are both especially important features in the context of environmental preservation.

Focusing on the solid plastic substrates for SPEs, recently, the reuse of PET from drink bottles as the substrate for the screen printing of disposable SPEs has been

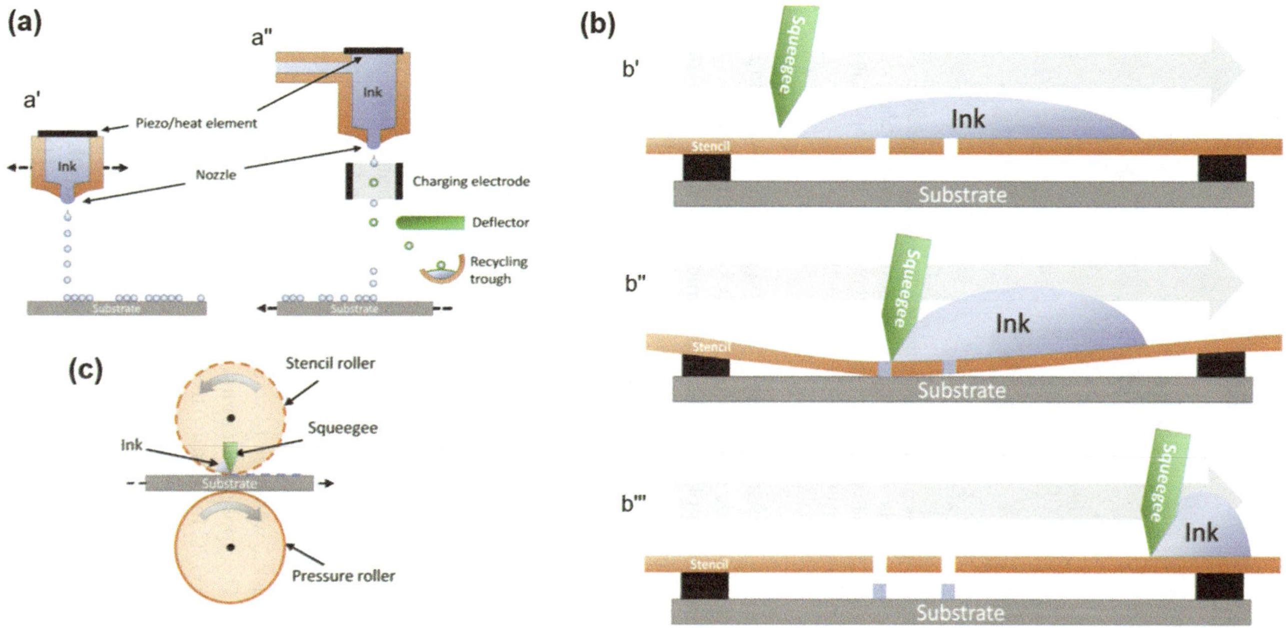

Fig. 4 Illustration of **a** inkjet printing (drop-on-demand (a′) and continuous (a″)); **b** screen-printing process: (b′) printing ink on the stencil is moved across the stencil with a squeegee and (b″) deposited through voids in the stencil, (b‴) after the stencil is lifted off, the printed pattern remains on the substrate; **c** roll-to-roll screen printing. Reprinted from [12] with permission of Elsevier

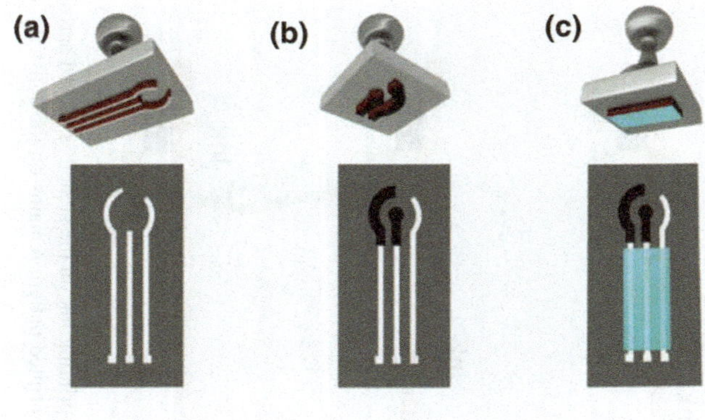

Fig. 5 Illustration detailing the steps involved in the stamp transfer preparation of a common electrode pattern using elastomeric stamps with custom-designed surface reliefs containing the electrode pattern (top): **a** Ag/AgCl conductor layer, **b** carbon active layer, and **c** insulator layer. Reproduced from [14] with permission from the Royal Society of Chemistry

reported. An alternative low-cost carbon ink based on graphite powder and nail polish [15] has been used for electrode preparation. The flexible SPEs manufactured by this method were successfully applied at the voltammetric sensing of hydroquinone, epinephrine, and serotonin. Another alternative strategy for SPEs preparation on plastic substrates was recently developed by Faria and collaborators [16]. In this case, an all-plastic disposable electrochemical cell (DCell) was designed and manufactured using a home-cut printer. A template was prepared by cutting an adhesive vinyl, which is transfer to the transparency sheet (polyester). Electrodes are printed on substrate using carbon or Ag/AgCl ink. Figure 6 outlines the steps of the proposed methodology.

On the use of paper as a substrate for the direct printing of SPEs, Cinti et al. [17] investigated the manufacturing of SPEs using three different types of paper normally not applied at electrochemistry fields. The evaluated substrates were paper towel, waxed paper, and Parafilm M®. Figure 7 provides pictures and the dimensions of the manufactured SPEs. The SPEs were manually screen printed using two masks (masks 1 and 2) and a squeegee. Using mask 1, connections and pseudo-reference electrode were printed; then, mask 2 was used to screen print the working (diameter of 4 mm) and counter electrodes by applying a commercial carbon ink; to finish, the SPEs were cured at mild heating (60 °C for 30 min) to prevent the damage of the paper substrate. In this context, the authors observed an apparent better performance of those SPEs based on towel paper, with an enhanced voltammetric response after the incorporation of cost-effective carbon black (CB) nanoparticles on the working electrode. In the context of SPEs manufactured on paper substrates, it is worth commenting on a recent trend related to the use of pencil/pen drawn electrodes [18]. In this case, the electrodes are manually drawn using commercially available graphite pencils [19], or ballpoint pens are adapted to directly dispense the conductive ink on the paper

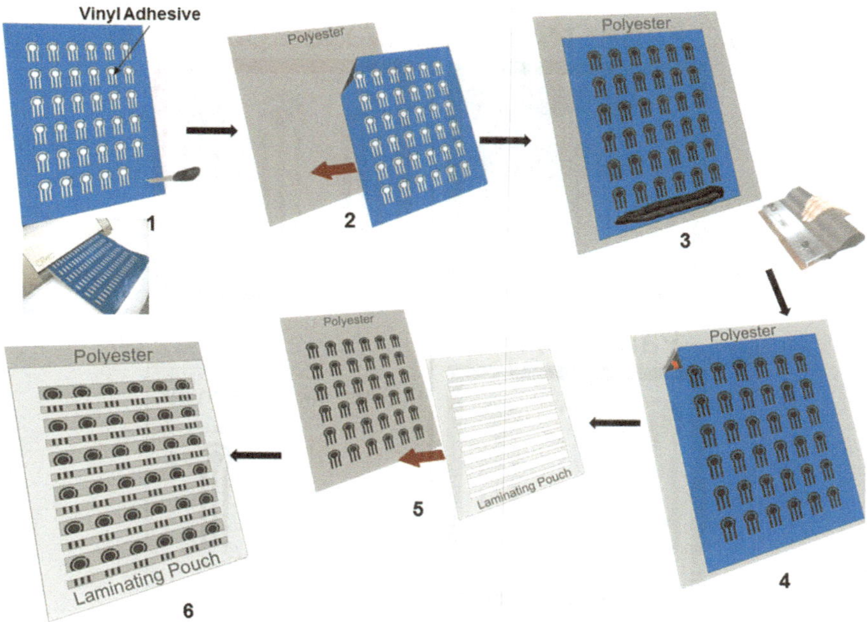

Fig. 6 Schematic representation of DCellf fabrication: (1) preparation of the template by cutting the adhesive vinyl; (2) transfer of the vinyl to the transparency sheet; (3) carbon ink screen printing, followed by Ag/AgCl ink deposition; (4) removal of the adhesive vinyl; (5) positioning of the lamination pouch on the printed electrodes, following by heating and pressing; and (6) the DCell ready for use. Reprinted from [16] with permission of Elsevier

surface [20]. Toner lines or stencils can be used to delimit the geometry and areas of the electrodes [18]. The simplicity of these printing methods is impressive; however, the fact that they are performed manually can compromise the reproducibility of manufacturing, which would difficult to scale up them for mass SPE production [18].

In addition to the substrate, conductive ink is a crucial component in making SPEs. The conductive inks applied are those classically used in various electronic components, consisting of three main chemical components: conductive material, binder, and solvent [21]. Each of these components aims to meet specific prerequisites for the desired application, such as high electrical conductivity, ink homogeneity, and the formation of well-adherent coatings on the substrates of interest. Regarding the binder agent, several surfactants and polymers have been used and act as a stabilizer for the ink [21]. Organic solvents (e.g., toluene, isophorone, acetone, etc.) provide important advantages as ease of application, low viscosity, and fast drying. However, the toxicity and high costs associated with the organic solvents make it difficult to use them on a large scale, and efforts have been expended on the formulation of eco-friendly aqueous inks [22]. In the case of the conductive material, these can be carbon-based material and/or metallic species. Special emphasis can be placed

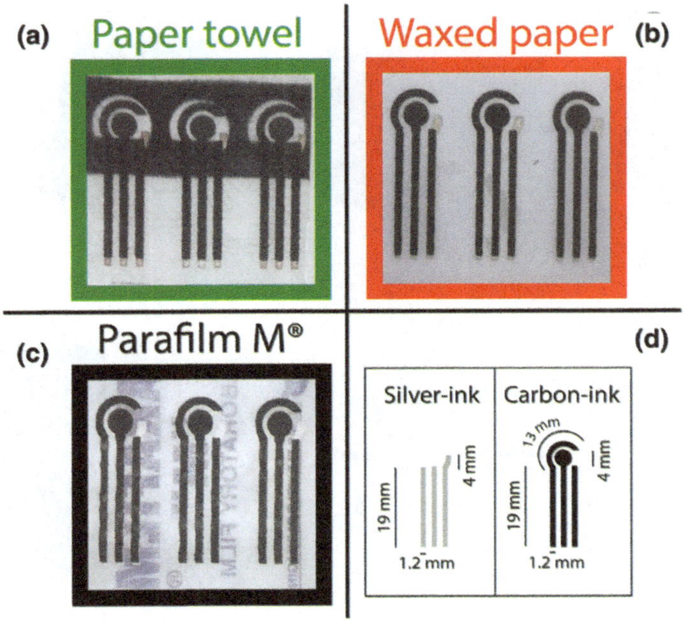

Fig. 7 Photographs of electrodes screen printed on **a** paper towel; **b** waxed paper; and **c** Parafilm®; **d** dimensions of the conductive tracks that have been screen printed. Reprinted from [17] with permission of Royal Society of Chemistry

on graphite-based carbon conductive inks due to adequate electrical and thermal conductivities, wide availability, and the low cost of this material. Graphite-based SPEs, due to the presence of polymers as the ink binders, are often insulating, and it could be necessary to carry out chemical and/or electrochemical pre-treatments to activate the electrode surface from the removal of the polymer matrix at the interfacial region (electrode/solution) [23]. Other carbonaceous materials are also explored for the formulation of conductive inks besides graphite, such as the conductive inks based on graphene [24], carbon nanotubes [25], and carbon black [26, 27]. In addition to carbon-based conductive inks, there are also conductive inks consisting on dispersions of noble metal nanoparticles (mainly AuNPs [28], AgNPs [29], PtNPs [30], and CuNPs [31]). AgNPs-based inks are most widely used, especially in the preparation of pseudo-reference electrodes; however, in fact, one sees a preference for carbonaceous conductive inks due to the lower production cost. An important aspect of the preparation of conductive inks is their homogeneity. To ensure this aspect and the reproducibility of electrode construction, protocols involving ultrasonication [32] or double asymmetric centrifugation [33] have been implemented. Practical details and perspectives on preparing conductive inks for printed electrochemical devices can be found in our very recently published review article [21].

SPEs can be applied directly or undergo a preliminary modification step. This modification can take place in the manufacturing process by incorporating the modifier in the conductive ink formulation or as a further modification step by drop casting, dip coating, or spin coating of a solvent-modifier mixture, or even via electrodeposition, electrospray, and electrospun [34, 35]. Nanostructured materials are well explored to improve the analytical response of sensors and biosensors based on SPEs, highlighting the use of carbonaceous nanomaterials (carbon black, carbon nanotubes, graphene, among others), ionic liquids, metallic and metallic oxide nanoparticles, magnetic nanoparticles, and conductive polymers [34]. The modification of the surface of SPE has gained attention since it is the feasible platform for the preparation of electrochemical sensors with low-cost and high analytical performance. Among several modifiers, the use of biological species such as enzymes, nucleic acids, and antibodies/antigens allowed the development of enzymatic biosensors, genosensors, and immunosensors for the determination of important biomarkers. The biological recognition site can be immobilized onto the surface of the WE through different physical (e.g., adsorption and entrapment) and chemical (e.g., crosslinking and covalent attachment) methods [11].

2 Electrochemical Determination of Biomarkers Using 2D-Printed Devices

Throughout this section, the recent scientific advances in the use of electrochemical devices obtained by 2D printing will be discussed. The number of contributions in the literature is quite significant; therefore, it is not the intention to exhaust the literature, but to point out promising results from the last years, bringing new perspectives to this important field of research. In Table 1, the reviewed papers dedicated to the development and application of 2D electrochemical devices for the determination of biomarkers are organized. Featuring by the presence on the sensing surface of a biorecognition agent, the electrochemical biosensors currently tend to be preferred when it comes to detecting biomarkers due to their superior selectivity and specificity. An overview of this table shows the applicability of 2D-printed electrodes for the incorporation of different types of recognition agents (enzymes, DNA, and antibodies/antigens) in addition to various nanostructured architectures, for the voltammetric, (chrono)amperometric, impedimetric, and/or coulometric detection of biomarkers related to various diseases (e.g., cancers, Alzheimer's, and diabetes). The main sample matrix subject to electrochemical analysis is human serum samples from healthy and sick patients. The recent researches use commercial-printed or laboratory-manufactured electrodes with classic formats to more complex electrode arrays in microfluidic systems, allowing the multiplexed determination of biomarkers and including the proposition of flexible and wearable devices coupled to portable potentiostats equipped with a system of real-time data transmission. These general

Table 1 Analytical features of 2D electrochemical (bio)sensors toward biomarkers determination

Biomarker	Electrode	Technique	Concentration range	LOD	Sample	References
$A\beta_{1\text{-}40}$	rGO SPEs	DPV	10 fmol L^{-1}–10 pmol L^{-1}	9.51 fmol L^{-1}	Plasma	[43]
$A\beta_{1\text{-}42}$	rGO SPEs	DPV	10 fmol L^{-1}–50 pmol L^{-1}	8.65 fmol L^{-1}	Plasma	[43]
NfL	MBs-SCPEs	Amperometry	10.0 pg L^{-1}–5000.0 pg L^{-1}	3.0 pg mL^{-1}	Plasma and brain tissue	[44]
EpCAM positive cancer cells	Apt/GNST/SiNP-IL/SPE	DPV	5–10^7 cells mL^{-1}	1 cell mL^{-1}	Serum	[45]
Pyocyanin	AuNPs/rGO/SPCE	DPV	1.0 µmol L^{-1}–100.0 pmol L^{-1}	0.27 µmol L^{-1}	*P. aeruginosa* culture	[46]
Her-2	LSG-AuNS	SWV	0.1–200.0 ng mL^{-1}	0.008 ng mL^{-1}	Serum	[47]
Leptin	oPD/Streptavidin-HRP/Biotinylated α-Leptin/α-leptin/DTSSP/SPGEs	SWV	0.1–20.0 ng mL^{-1}	0.033 ng mL^{-1}	Mouse blood serum	[48]
IgG	AgNPs/rGO/SPCE	SWV	1.0–50.0 pg mL^{-1} and 0.05–50.0 ng mL^{-1}	0.086 pg mL^{-1}	Serum	[37]
miR-21	ANSAm-DNA/SPCE	Coulometry	$1.0 \times 10^{-14 \text{ mol}}$ L^{-1}–$1.0 \times 10^{-8 \text{ mol}}$ L^{-1}	17.0 fmol L^{-1}	Urine	[36]
circRNA	AuNPs-SPME	DPV	1.0 pmol L^{-1}–100.0 nmol L^{-1}	10.0 pmol L^{-1}	Serum	[49]
PD-L1	SPCE	Amperometry	290–2500.0 pg mL^{-1}	86 pg mL^{-1}	Cancer cell lysates	[50]
HIF-1α	SPCE	Amperometry	930–10,000 pg mL^{-1}	279 pg mL^{-1}	Cancer cell lysates	[50]
lncRNA biomarker MALAT1	Au NCs/MWCNT-NH_2-SPCE	DPV	$1.0 \times 10^{-14 \text{ mol}}$ L^{-1}–$1.0 \times 10^{-7 \text{ mol}}$ L^{-1}	48.0 fmol L^{-1}	Serum	[51]
miR-141	HRP-modified SPCE	Chronoamperometry	$1.0 \times 10^{-12 \text{ mol}}$ L^{-1}–$1.0 \times 10^{-8 \text{ mol}}$ L^{-1}	0.1 pmol L^{-1}	Cohort of colorectal and breast cancer samples	[52]

(continued)

Table 1 (continued)

Biomarker	Electrode	Technique	Concentration range	LOD	Sample	References
GPC3	RGO–H–CS/SPE	DPV	0.01–10.0 μg mL^{-1}	7.9 ng mL^{-1}	Serum	[53]
miRNA-21	AuNPs/GQDs/GO/3SPCE	SWV	0.01–1000 pmol L^{-1}	0.04 fmol L^{-1}	Serum	[54]
miRNA-155	AuNPs/GQDs/GO/3SPCE	SWV	0.01–1000 pmol L^{-1}	0.33 fmol L^{-1}	Serum	[54]
miRNA-210	AuNPs/GQDs/GO/3SPCE	SWV	0.01–1000 pmol L^{-1}	0.28 fmol L^{-1}	Serum	[54]
cTnl	pAb_2-PBDENP/cTnl/mAb_1-BSA-SPCE	SWV	0.01–100 ng mL^{-1}	6.2 pg mL^{-1}	Serum	[55]
NIM	Fe_3O_4-3D GPE	SWV	0.8–6 μmol L^{-1}	0.01 μmol L^{-1}	Serum and artificial urine	[56]
DOP	Fe_3O_4-3D GPE	SWV	0.5–9 μmol L^{-1}	0.0023 μmol L^{-1}	Serum and artificial urine	[56]
UA	Fe_3O_4-3D GPE	SWV	0.7–10 μmol L^{-1}	0.0034 μmol L^{-1}	Serum and artificial urine	[56]
Creatinine	Nafion/polyacrylic gel–Cu^{2+}/cuprous oxide NPs/SPCE	CV/DPV/impedance/IoT	1–2000 μmol L^{-1}	0.3 μmol L^{-1}	Artificial and in human saliva	[57]
p53	AuNP-SPCE	LSV	2 nmol L^{-1}–50 mol L^{-1}	0.05 nmol L^{-1}	Plasma	[58]
Uric acid	SPCE-MNC	DPV	0–70 μmol L^{-1}	1.8 μmol L^{-1}	Artificial sweat	[59]
17β-estradiol	SPCE-MNC	DPV	0–3 μmol L^{-1}	0.27 μmol L^{-1}	Artificial sweat	[59]
NGAL	Dual-SPCE	LSV	0.15–2.0 ng mL^{-1}	0.096 ng mL^{-1}	Urine	[60]
HER2	HOOC-MBs/SPCE	LSV	5.0–50 ng $mL^{-1}$and 50–100 ng mL^{-1}	2.8 ng mL^{-1}	Serum	[61]
miR-21	rGO/Au/SPE	DPV	$1.0 \times 10^{-12\,mol}$ L^{-1}–$1.0 \times 10^{-4\,mol}$ L^{-1}	1.0 pmol L^{-1}	Saliva	[62]
RANKL	Neu-MBs/SPdCEs	Amperometry	0.0–1000 pg mL^{-1}	2.6 pg mL^{-1}	Serum	[63]
TNF	Neu-MBs/SPdCEs	Amperometry	0.0–1000 pg mL^{-1}	3.0 pg mL^{-1}	Serum	[63]
Tau protein	3D-Au-PAMAM-p-ABA-SPCE	Amperometry	6.0–5,000 pg mL^{-1}	1.7 pg mL^{-1}	Plasma and brain tissue extracts	[64]

(continued)

Table 1 (continued)

Biomarker	Electrode	Technique	Concentration range	LOD	Sample	References
HOTAIR RNA	SPE-Au	Chronoamperometry	1 fmol L^{-1}–1 nmol L^{-1}	1.0 fmol L^{-1}	Ovarian cancer cell lines and plasma	[65]
Gal-3	SPCE	Chronoamperometry	2.8 × $10^{-2\ ng\ mL-1}$–5 ng mL^{-1}	8.3 pg mL^{-1}	Plasma	[66]
PSA	ce-MoS_2/AgNR/SPE	CV	0.1–1000 ng mL^{-1}	0.051 ng mL^{-1}	–	[67]
p53	SPCEs/PEI/NPs	DPV	1.0–10 × $10^{3\ pg\ mL-1}$	5.0 fg mL^{-1}	Fetal bovine serum, cell lysate, and saliva	[68]
CA19-9	SPCE/CB-PEI/CB-PAA	DPV	0.01–40 U mL^{-1}	0.07 U mL^{-1}	Serum and cell lysate	[69]
serpin A12	MCH/thiolated aptamer/FLGMs/SPCE	DPV	0.039 ng mL^{-1}–10 ng mL^{-1}	0.02 ng mL^{-1}	Plasma	[70]
miR-21	His-Tag-Isolation-MBs/SPCE	Amperometry	3.0–100 nmol L^{-1}	0.91 nmol L^{-1}	RNA_t extracted from epithelial non-tumorigenic and adenocarcinoma breast cells	[71]
$A\beta_{1-42}$	Au film graphene/rGO-SPE	DPV	11 pmol L^{-1}–55 ^{nmol} L^{-1}	2.398 pmol L^{-1}	Human and mice plasmas	[72]
Insulin	MWNT-QD modified SPE array	SWV	100–5000 pmol L^{-1}	100 pmol L^{-1}	– -	[73]
CYFRA 21–1	CoZnFeONPs	Amperometry	3.9–1000 fg mL^{-1}	0.19 fg mL^{-1}	Serum samples from healthy and prostate cancer patients	[37]
PSA	MIP-SPE	EIS	0.01–100 ng mL^{-1}	5.4 pg mL^{-1}	Serum and urine	[74]
Myo	MIP-SPE	EIS	1 nm $mL^{-1-20,000\ ng\ mL-1}$	0.83 ng mL^{-1}	Serum and urine	[74]

(continued)

Table 1 (continued)

Biomarker	Electrode	Technique	Concentration range	LOD	Sample	References
INO	$(PNP\text{-}CS)_2$-SPCE	SWV	2–90 μmol L^{-1}	0.3 μmol L^{-1}	Serum (from normal and myocardial infarction patients) and dietary supplements	[75]
CA19-9	SPIDE/CNO-GO	EIS	0.3–100 U mL^{-1}	0.12 U mL^{-1}	Whole-cell lysates of colorectal adenocarcinoma	[76]
CEA	NG-THI-AuNPs modified paper electrode	DPV	0.01–500 ng mL^{-1}	2.0 pg mL^{-1}	Serum	[77]
NSE	PB-PEDOT-AuNPs modified paper electrode	DPV	0.05–500 ng mL^{-1}	10.0 pg mL^{-1}	Serum	[77]
CXCL7	IgG-MWCNTs/SPCE	DPV	0.5–600 pg mL^{-1}	0.1 pg mL^{-1}	Serum samples from patients with rheumatoid arthritis	[78]
IL-13Rα2	SPdCEs	Amperometry	3.4–100 ng mL^{-1}	1.03 ng mL^{-1}	Serum and paraffined-embedded tissues from colorectal cancer (CRC) patients	[79]
E-CDH	SPdCEs	Amperometry	0.9–25 ng mL^{-1}	3.45 ng mL^{-1}	Serum and paraffined-embedded tissues from colorectal cancer (CRC) patients	[79]
AFP	PEI-AuNPs/SPCE	SWV	0.25–10 ng mL^{-1}	1.7 fg mL^{-1}	Serum	[80]
CEA	PEI-AuNPs/SPCE	SWV	0.25–10 ng mL^{-1}	1.6 fg mL^{-1}	Serum	[80]
PSA	PEI-AuNPs/SPCE	SWV	0.25–10 ng mL^{-1}	0.9 fg mL^{-1}	Serum	[80]

(continued)

Table 1 (continued)

Biomarker	Electrode	Technique	Concentration range	LOD	Sample	References
IL-8	PEI-AuNPs/SPCE	SWV	0.50–100 pg mL^{-1}	1.0 fg mL^{-1}	Serum	[80]
PSA	CASPE-MFD	Chronoamperometry	0.001–10 ng mL^{-1}	0.84 Pg mL^{-1}	Serum	[81]
8-OHdG	SPE-Gr	LSV	$3.0 \times 10^{-7\,mol}$ L^{-1}–$1.0 \times 10^{-4\,mol}$ L^{-1}	90 nmol L^{-1}	Saliva	[82]
p24-HIV	GQD-SPE	CV	0.93 ng mL^{-1}–93 µg mL^{-1}	51.7 pg mL^{-1}	Serum	

$A\beta_{1\text{-}40}$ e $A\beta_{1\text{-}42}$ Alzheimer's disease biomarkers; *rGO* reduced graphene oxide; *NfL* neurofilament light chain, an Alzheimer's disease biomarker; *MBs* magnetic microbeads; *Apt* aptamer; *GNST* gold nanostar structures; *SiNP* porous silica nanoparticles; *IL* ionic liquid; *CTCs*; *DPV* differential pulse voltammetry; *SCPEs* screen-printed carbon electrodes; EID50; *Her-2* human epidermal growth factor receptor 2; *SWV* square-wave voltammetry; *LSG-AuNS* nanostructured gold modified laser-scribed graphene-based electrode; *oPD* o-phenylenediamine; *SPGEs* screen-printed gold electrodes; *HRP* horseradish peroxidase; *DTSSP* sulfosuccinimidyl propionate; *IgG* human immunoglobulin G; *miR-21* urinary microRNA; *ANSAm* 4-amino-3-hydroxy-1-napthalene sulfonic acid; *circRNA* circular RNA from hepatocellular cancer; *SPME* screen-printed magnetic electrode; *PD-L1* programmed death ligand 1 protein; *HIF-1α* hypoxia-inducible factor 1 alpha protein; *Au NCs/MWCNT-NH_2* gold nanocage coupled with an amidated multi-walled carbon nanotube; *lncRNA biomarker MALAT1* long non-coding RNA biomarker for the screening of non-small cell lung cancer; *miR-141* microRNA associated to cancers such as colorectal cancer (CRC) and breast cancer (BC); *GPC3* glypican-3; *cTnI* cardiac troponin I; *PBDENPs* Prussian blue-doped PAMAM dendrimer nanospheres; *3D GPE* graphene paste electrodes; *IoT* Internet of Things; *p53* protein biomarker for Alzheimer's disease; *LSV* linear sweep voltammetry; *NGAL* neutrophil gelatinase-associated lipocalin; *HER2* human epidermal growth factor receptor 2; *HOOC-MBs* biomodified carboxylic acid functionalized magnetic beads; *RANKL* receptor activator of nuclear factor-κB related to breast cancer (BC); *TNF* tumor necrosis factor alpha related to breast cancer (BC); *Neu-MBs* neutravidin-functionalized magnetic microbeads; *SPdCEs* dual screen-printed carbon electrodes; *3D-Au-PAMAM* gold nanoparticles-poly(amidoamine) dendrimer nanocomposite; *p-ABA* p-aminobenzoic acid; *Tau protein* hallmark of Alzheimer's disease; *HOTAIR RNA* long non-coding RNA HOX transcript antisense intergenic; *SPE-Au* screen-printed gold electrode; *Gal-3* galectin-3, a β-galactosidase-binding lectin that acts as mediator of heart failure (HF); *PSA* prostate-specific antigen; *ce-MoS_2* chemical exfoliated MoS_2 nanosheets; *AgNR* Ag nanorods; *CV* cyclic voltammetry; *NPs* carboxylated $NiFe_2O_4$ nanoparticles; *PEI* polyethyleneimine; *CA19-9* cancer marker carbohydrate antigen 19-9; *CB-PEI* carbon black-polyethyleneimine; *CB-PAA* carbon black-polyacrylic acid; *serpin A12* type 2 diabetes biomarker; *MCH* 6-mercapto-1-hexanol; *FLGMs* flower-like gold microstructures; *His-Tag-Isolation-MBs* commercial His-Tag-Zinc finger protein conjugated to magnetic beads; *RNA_t* total RNA; *$A\beta_{1\text{-}42}$* beta-amyloid biomarker in Alzheimer's disease; *MWNT-QD* quantum dot decorated multi-walled carbon nanotube; *CYFRA 21-1* protein biomarker in different types of cancer; *CoZnFeONPs* $Co_{0.25}Zn_{0.75}Fe_2O_4$ nanoparticles; *Myo* myoglobin; *EIS* electrochemical impedance spectroscopy; *MIP* molecularly imprinted polymer; *INO* inosine; *PNP* purine nucleoside phosphorylase; *CS* chitosan; *CA19-9* pancreatic cancer biomarker; *SPIDE* screen-printed interdigitated electrodes; *CNO* carbon nano-onion; *GO* graphene oxide; *NG-THI-AuNPs* amino functional graphene–thionin–gold nanoparticles; *PB-PEDOT-AuNPs* Prussian blue-poly (3,4-ethylenedioxythiophene)-gold nanoparticles; *CXCL7* chemokine (C-X-C motif) ligand autoimmune biomarker; *IL-13Rα2* interleukin-13 receptor α2; *E-CDH* E-cadherin; *AFP* alpha fetoprotein; *CEA* carcinoembryonic antigen; *IL-8* interleukin-8; *CASPE-MFD* screen-printed electrode-based microfluidic device; *8-OHdG* 8-hydroxy-2′-deoxyguanosine; *p24-HIV* protein monitored in the early diagnosis of HIV; *GQD* graphene quantum dots

first observations can be better visualized in the following lines, where some advances mentioned briefly are better detailed to the reader.

In the scenario of diabetes control, a novel electroanalytical approach to evaluate diabetic kidney disease (DKD) has been proposed by Smith et al. [36]. In this case, a three-working electrodes configuration of SPCE was adopted to design a disposable electrochemical biosensor for the detection of microRNA (specifically miR-21) DKD biomarker at patient urine samples. For the biosensor construction, the carbon-based working electrode surface was primarily modified from the electrochemical deposition of 4-amino-3-hydroxy-1-napthalene sulfonic acid (ANSA), which was then chemically converted into a sulfonyl chloride (ANSCl). After, a 5′-amine-tagged miRNA-specific DNA oligonucleotide was attached via a sulfonamide linkage to complete the biosensor construction. The miR-21 concentration was monitored through chronocoulometric measurements carried out with negative and positive potential sweeps in the presence of the ferri/ferrocyanide redox couple.

Illustrating the use of nano-modified SPEs at the design immunosensors for biomarkers detections, Chanarsa et al. [37] have recently proposed the use of a silver nanoparticles (AgNPs)-reduced graphene oxide (rGO) nanocomposite synthesized by the one-pot wet chemical process for both the SPE modification and analytical signal generation (label-free approach). The scheme of preparation of the biosensor is displayed in Fig. 8. Thus, briefly, the working electrode surface of the SPCE is first modified by drop casting using an AgNPs-rGO dispersion, followed by incubation periods, and washing steps with PBS buffer to remove unbonded molecules, to sequentially incorporate onto AgNPs-rGO modified electrode anti-IgG antibodies, block non-specific binding sites (BSA) and target IgG model biomarker. The typical anodic peak for AgNPs, corresponding to the Ag^0 oxidation ($Ag^0 \rightarrow Ag^+ + e^-$), was explored to the label-free monitoring of IgG model biomarker. As the demonstrative

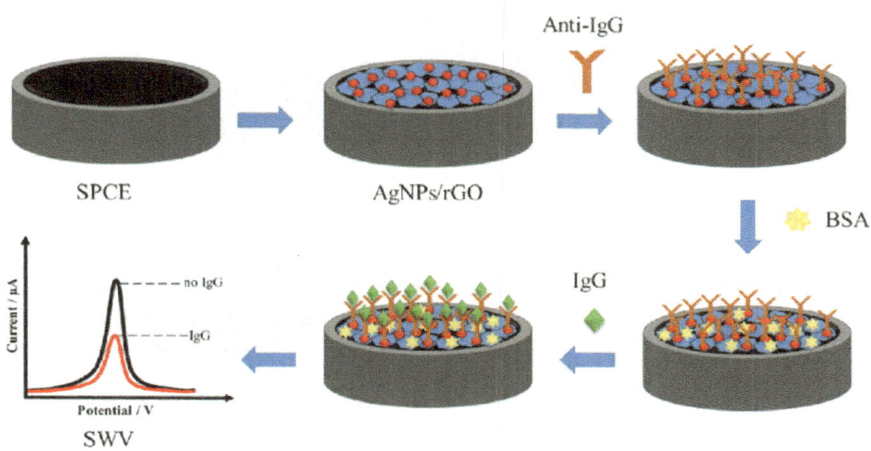

Fig. 8 Fabrication of AgNP/rGO-based electrochemical immunosensor. Reprinted from Ref. [37] with permission of Frontiers

square-wave voltammograms (SWV) of Fig. 8 suggests, the anodic peak current of the AgNPs process suffered deflection in the presence of the analyte, and this signal drop is proportional to the biomarker concentration. Using this strategy, the label-free biosensor provided a low limit of detection (LOD), ensuring the detection of IgG in human serum samples.

Besides using three or dual screen-printed electrodes, indeed, the application of 2D-printed (bio)sensors consisting of multiple working electrodes coupled to microfluid paths is an important and current trend. In this scenario, some interesting achievements were reported by Proença et al. [38], who proposed a disposable microfluidic immunoarray device (μID) based on $Co_{0.25}Zn_{0.75}Fe_2O_4$ nanoparticles (CoZnFeONPs) as peroxidase mimics for the early diagnosis of cancer from the detection of CYFRA 21-1 biomarker. Figure 9 shows a picture from the disposable microfluidic system composed of SPCEs with eight working electrodes, one counter electrode, and one Ag/AgCl pseudo-reference electrode. Additionally, in practical terms, the method was able to detect CYFRA 21-1 in real serum samples of healthy and prostate cancer patients, with good correlation with results obtained from an enzyme-linked immunosorbent assay (ELISA).

In another recent application, the early diagnosis of Alzheimer's disease (AD) was reported by Oliveira et al. [39], where the immunomagnetic capture of the target protein (ADAM10 protein, as a potential AD biomarker) and separation from the

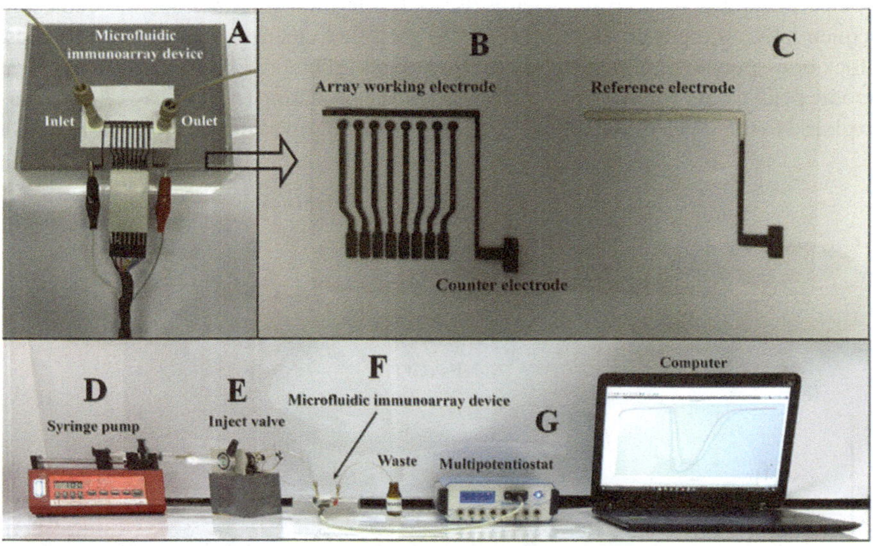

Fig. 9 Pictures of the disposable microfluidic device and the setup of the system: **a** disposable microfluidic immunoarray; **b** eight working electrodes and auxiliary electrode; and **c** reference electrode. The microfluidic system was composed of **d** syringe pump, **e** manual injection valve, **f** disposable microfluidic immunoarray, and **g** multipotentiostat. All parts were connected by using PEEK tubbing. Reprinted from [38] with permission of Elsevier

biological sample was employed as a part of the established biosensing procedure. Figure 10 outlines the involved steps, including the functionalization of magnetic beads (MB) with horseradish peroxidase (HRP) as a label and anti-ADAM10 antibodies (Fig. 10a), ADAM10 immunomagnetic capture in the sample, and injection of pre-treated samples in the microfluidic system coupled to an electrochemical detector.

Still in the scenario of 2D-printed electrochemical devices with multiple electrodes, suitable for multiplexed analyses, Fava et al. [40] have proposed a disposable microfluidic electrochemical paper-based device based on sixteen independent microfluidic channels with electrochemical detection. The design of the parts of the device containing the working electrodes, counter electrodes, and pseudo-reference electrodes is presented in Fig. 11a, b. The assembly of the device and a picture from the final device can be seen in Fig. 11c, d. Once again, the printing cut process was used to obtain the formats and masks for applying the carbon conductive ink. With this device, a small aliquot of the sample (60 μL) is injected into the central hole of

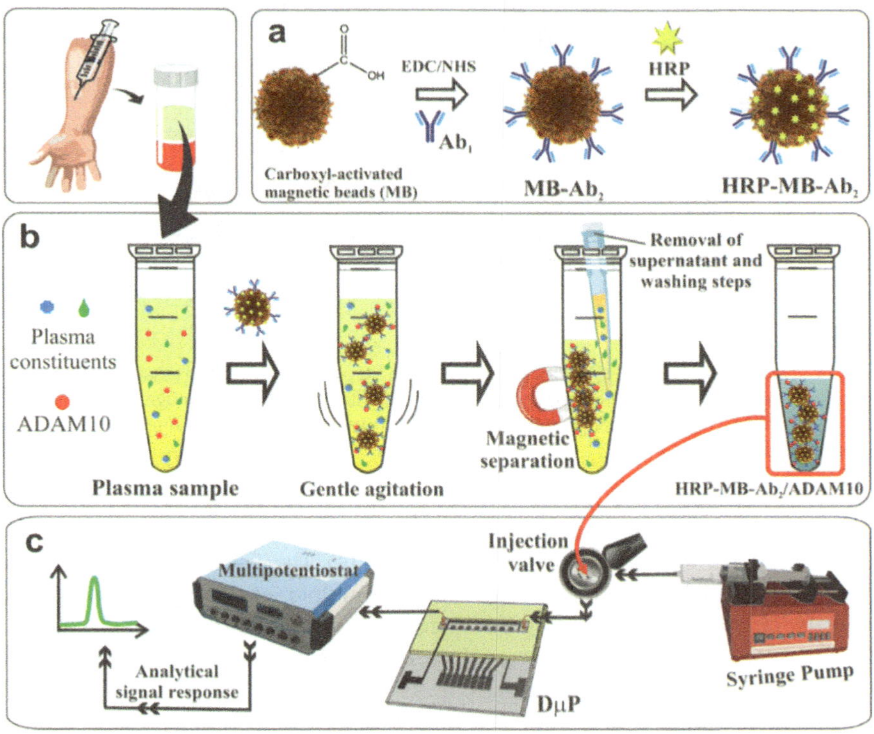

Fig. 10 Immunomagnetic capture of ADAM10 in a plasma sample and the detection using DμP. **a** Conjugation of the MBs with HRP and Ab_2. **b** ADAM10 immunomagnetic capture in the sample. **c** Setup of the microfluidic system and the transient current signal. Reprinted with permission from [39], Copyright (2020), American Chemical Society

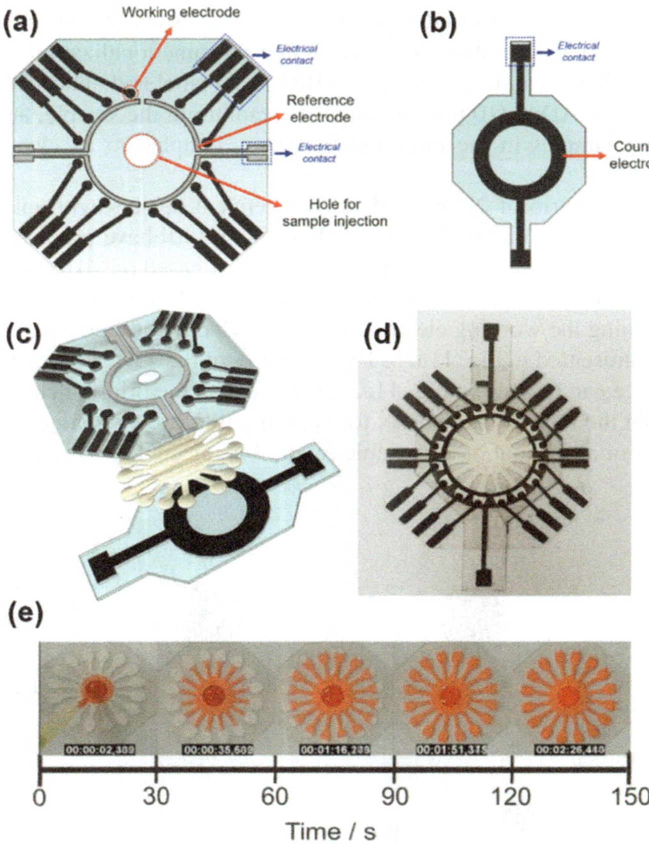

Fig. 11 Design of the **a** WEs/REs and **b** CE layers. **c** Steps for device layer assembly. **d** Photography of assembled microfluidic paper-based electrochemical device. **e** Homogeneous elution of $1.0 \times 10^{-3\,\mathrm{mol}}\ \mathrm{L}^{-1}$ Allura red dye solution in the microfluidic pathway. Reprinted from [40] with permission of Elsevier

the device, which flows through the paper microfluidic channels until reaching the 16 cells (sensing spots) circularly arranged (Fig. 11e). The working electrodes can be functionalized in the most varied ways to obtain multiple sensors and biosensors. For now, the modification of working electrodes with carbon black (CB) nanoparticles and the determination of biomarkers (glucose, creatinine, and uric acid) in real urine samples have been reported [41].

To produce methods that are less invasive and generate real-time results, 2D-printing technologies have provided important contributions to the field of wearable (bio)sensors. Several approaches can be found in the literature, such as the technology recently reported by Vinoth et al. [42]. The authors developed a wearable microfluidic sensor using a screen-printed carbon master for the multiplexed analysis of biomarkers at sweat during exercise activities (Fig. 12). The sweat sampling and sensing steps are taken in a low-dimensional microdevice containing silane functionalized microchannels, which enhance the sweat collecting at the inlet and the directing of samples to the surface of the printed sensor. For the electrochemical transduction (amperometry and potentiometry), a custom-made miniature printed circuit board (PCB) powered by a 200 mAh Li-ion rechargeable battery that enables multiplexed decoding of sweat and wireless signal transduction to the host device was applied. The developed system was mounted on the epidermis and evaluated toward the real-time multiplexed determination of lactate (via amperometric transduction) and Na^+, K^+, and pH (via potentiometric transduction) during stationary biking.

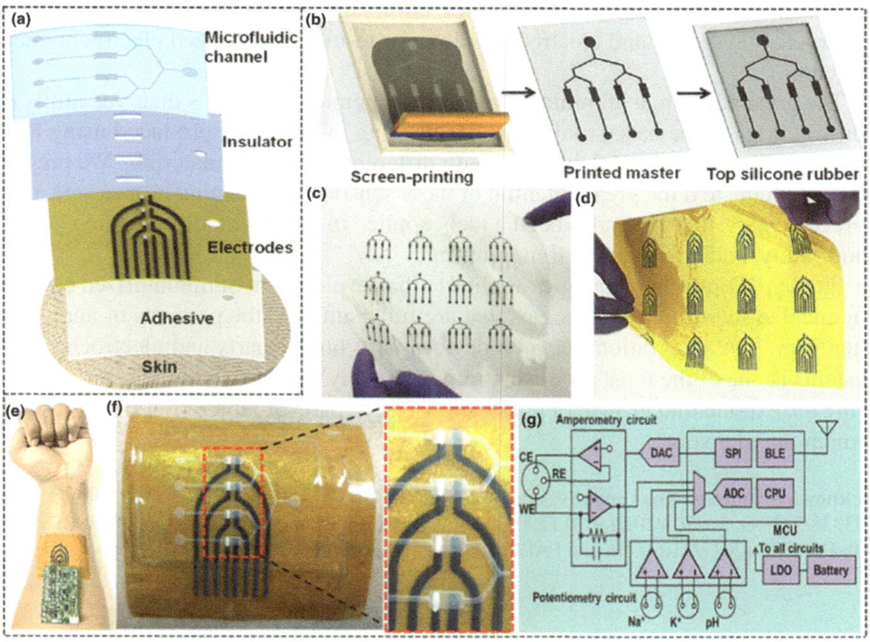

Fig. 12 Overview of the fully printed microfluidic device proposed by Vinoth and colleagues. **a** Schematic representation of different layers of the microfluidic patch. **b** Schematic illustration of the fabrication of silicone rubber-based microfluidic channels from the screen-printed carbon master. The optical images showing an array of printed **c** carbon master and **d** multiarray electrodes. Images showing **e** the skin-interfaced, fully integrated microfluidic sensor and **f** the close-up part of multiarray electrode having four sensing chambers. **g** Scheme representing the layout of miniature PCB capable of multiplexed analysis and wireless data transmission. Reprinted with permission from [42], Copyright (2020), American Chemical Society

3 Conclusions and Perspectives

Most biomarker analysis methods are based on laboratory analytical techniques that commonly require expensive and bulky equipment, including chromatographic or spectrometric methods, among others. In this sense, miniaturized analytical systems present special advantages such as portability and low cost and can be used for applications outside the laboratory environment to overcome the disadvantages of conventional analytical procedures, reduce total analytical time, or increase efficiency.

As presented in this book chapter, miniaturized devices based on electrochemical systems have evolved rapidly on monitoring of biomarkers due to several aspects as simplicity, high sensitivity and selectivity, reduced power requirements, and good compatibility with biological samples. In addition, its compatibility with advanced microfabrication technologies allows the replacement of conventional electrochemical cells and electrodes with easy-to-use miniaturized electrochemical systems.

The great advantage of miniaturized electrochemical devices is their versatility of construction. Those based mainly on 2D-printing methods are produced using low-cost techniques, easily scalable, and with different recyclable materials. We present along with the text the great potential of these sensors for detecting biomarkers, and, in recent years, they have also been widely applied in various fields, including health, food safety, and environmental monitoring.

Finally, 2D-printing strategies applied to the development of miniaturized electrochemical systems are technologies that are fully suited to the precepts of analytical chemistry. The integration between devices built on 2D parts and electrochemical systems is one of the most promising trends in today's electroanalytical, and considering the exponential growth of publications, these parts of fabrication strategies will remain popular soon.

Acknowledgements The authors are grateful to CNPq (402943/2016-3, 408309/2018-0, 303338/2019-9, 309803/2020-9, 311290/2020-5), CAPES (001 and Pandemias 8887.504861/2020-00), FAPESP (2017/21097-3), and FAPEMIG (APQ-00083-21) for the financial support.

References

1. Navarrete F, Sala F, Garc MS (2020) Biomarkers in psychiatry: concept, definition, types and relevance to the clinical reality, front. Psychiatry 11:1–14. https://doi.org/10.3389/fpsyt.2020.00432
2. Pletcher MJ, Pignone M (2011) Evaluating the clinical utility of a biomarker: a review of methods for estimating health impact. Circulation 123:1116–1124. https://doi.org/10.1161/CIRCULATIONAHA.110.943860.Evaluating
3. Aronson JK, Ferner RE (2017) Biomarkers—a general review. Curr Protocol Pharmacol 2017:9.23.1–9.23.17. https://doi.org/10.1002/cpph.19

4. Strimbu K, Tavel JA (2010) What are biomarkers? Curr Opin HIV AIDS 5:463–466. https://doi.org/10.1097/COH.0b013e32833ed177
5. Byrnes SA, Weigl BH (2018) Selecting analytical biomarkers for diagnostic applications: a first principles approach. Expert Rev Mol Diagn 18:19–26. https://doi.org/10.1080/14737159.2018.1412258
6. Reyes DR, Iossifidis D, Auroux PA, Manz A (2002) Micro total analysis systems. 1. Introduction, theory, and technology. Anal Chem 74:2623–2636. https://doi.org/10.1021/ac0202435
7. Ríos Á, Zougagh M, Avila M (2012) Miniaturization through lab-on-a-chip: Utopia or reality for routine laboratories? A review. Anal Chim Acta 740:1–11. https://doi.org/10.1016/j.aca.2012.06.024
8. Foster CW, Kadara RO, Banks CE (2016) Screen-printing electrochemical architectures
9. Tuteja SK, Ormsby C, Neethirajan S (2018) Noninvasive label-free detection of cortisol and lactate using graphene embedded screen-printed electrode. Nano-Micro Lett 10:1–10. https://doi.org/10.1007/s40820-018-0193-5
10. Nordin AN, Zainuddin AA, Rahim RA, Voiculescu I, Mak WC (2017) Screen printed electromechanical micro-total analysis system (μtas) for sensitive and rapid detection of infectious diseases. Procedia Technol 27:100–101. https://doi.org/10.1016/J.PROTCY.2017.04.043
11. Couto RAS, Lima JLFC, Quinaz MB (2016) Recent developments, characteristics and potential applications of screen-printed electrodes in pharmaceutical and biological analysis. Talanta 146:801–814. https://doi.org/10.1016/j.talanta.2015.06.011
12. Mensing JP, Lomas T, Tuantranont A (2020) 2D and 3D printing for graphene based supercapacitors and batteries: a review. Sustain Mater Technol 25:e00190. https://doi.org/10.1016/j.susmat.2020.e00190
13. Hu G, Kang J, Ng LWT, Zhu X, Howe RCT, Jones CG, Hersam MC, Hasan T (2018) Functional inks and printing of two-dimensional materials. Chem Soc Rev 47:3265–3300. https://doi.org/10.1039/C8CS00084K
14. Windmiller JR, Bandodkar AJ, Parkhomovsky S, Wang J (2012) Stamp transfer electrodes for electrochemical sensing on non-planar and oversized surfaces. Analyst 137:1570–1575. https://doi.org/10.1039/C2AN35041F
15. de Araujo Andreotti IA, Orzari LO, Camargo JR, Faria RC, Marcolino-Junior LH, Bergamini MF, Gatti A, Janegitz BC (2019) Disposable and flexible electrochemical sensor made by recyclable material and low cost conductive ink. J Electroanal Chem 840:109–116. https://doi.org/10.1016/j.jelechem.2019.03.059
16. Afonso AS, Uliana CV, Martucci DH, Faria RC (2016) Simple and rapid fabrication of disposable carbon-based electrochemical cells using an electronic craft cutter for sensor and biosensor applications. Talanta 146:381–387. https://doi.org/10.1016/j.talanta.2015.09.002
17. Cinti S, Mazzaracchio V, Cacciotti I, Moscone D, Arduini F (2017) Carbon black-modified electrodes screen-printed onto paper towel, waxed paper and Parafilm M®. Sensors 17. https://doi.org/10.3390/s17102267
18. Noviana E, McCord CP, Clark KM, Jang I, Henry CS (2020) Electrochemical paper-based devices: sensing approaches and progress toward practical applications. Lab Chip 20:9–34. https://doi.org/10.1039/C9LC00903E
19. Dias AA, Cardoso TMG, Chagas CLS, Oliveira VXG, Munoz RAA, Henry CS, Santana MHP, Paixão TRLC, Coltro WKT (2018) Detection of analgesics and sedation drugs in whiskey using electrochemical paper-based analytical devices. Electroanalysis 30:2250–2257. https://doi.org/10.1002/elan.201800308
20. Yi TF, Sari HMK, Li X, Wang F, Zhu YR, Hu J, Zhang J, Li X (2021) A review of niobium oxides based nanocomposites for lithium-ion batteries, sodium-ion batteries and supercapacitors. Nano Energy 85:105955. https://doi.org/10.1016/j.nanoen.2021.105955
21. Camargo JR, Orzari LO, Araújo DAG, de Oliveira PR, Kalinke C, Rocha DP, Luiz dos Santos A, Takeuchi RM, Munoz RAA, Bonacin JA, Janegitz BC (2021) Development of conductive inks for electrochemical sensors and biosensors. Microchem J 164:105998. https://doi.org/10.1016/j.microc.2021.105998

22. Parvez K, Worsley R, Alieva A, Felten A, Casiraghi C (2019) Water-based and inkjet printable inks made by electrochemically exfoliated grapheme. Carbon N Y 149:213–221. https://doi.org/10.1016/j.carbon.2019.04.047
23. Wei H, Sun J-J, Xie Y, Lin C-G, Wang Y-M, Yin W-H, Chen G-N (2007) Enhanced electrochemical performance at screen-printed carbon electrodes by a new pretreating procedure. Anal Chim Acta 588:297–303. https://doi.org/10.1016/j.aca.2007.02.006
24. Labroo P, Cui Y (2014) Graphene nano-ink biosensor arrays on a microfluidic paper for multiplexed detection of metabolites. Anal Chim Acta 813:90–96. https://doi.org/10.1016/j.aca.2014.01.024
25. Han J-W, Kim B, Li J, Meyyappan M (2014) Carbon nanotube ink for writing on cellulose paper. Mater Res Bull 50:249–253. https://doi.org/10.1016/j.materresbull.2013.10.048
26. Sekertekin Y, Bozyel I, Gokcen D (2020) A flexible and low-cost tactile sensor produced by screen printing of carbon black/PVA composite on cellulose paper. Sensors 20. https://doi.org/10.3390/s20102908
27. dos Santos SMV, Oliveira PR, Oliveira MC, Bergamini MF, Marcolino-Jr LH (2017) Eletrodos impressos construídos por serigrafia utilizando negro de fumo como material conductor. Rev Virtual Química 9
28. Pavithra M, Muruganand S, Parthiban C (2018) Development of novel paper based electrochemical immunosensor with self-made gold nanoparticle ink and quinone derivate for highly sensitive carcinoembryonic antigen. Sens Actuators B Chem 257:496–503. https://doi.org/10.1016/j.snb.2017.10.177
29. Fernandes IJ, Aroche AF, Schuck A, Lamberty P, Peter CR, Hasenkamp W, Rocha TLAC (2020) Silver nanoparticle conductive inks: synthesis, characterization, and fabrication of inkjet-printed flexible electrodes. Sci Rep 10:8878. https://doi.org/10.1038/s41598-020-65698-3
30. Abellán-Llobregat A, Jeerapan I, Bandodkar A, Vidal L, Canals A, Wang J, Morallón E (2017) A stretchable and screen-printed electrochemical sensor for glucose determination in human perspiration. Biosens Bioelectron 91:885–891. https://doi.org/10.1016/j.bios.2017.01.058
31. Li W, Chen M (2014) Synthesis of stable ultra-small Cu nanoparticles for direct writing flexible electronics. Appl Surf Sci 290:240–245. https://doi.org/10.1016/j.apsusc.2013.11.057
32. Bi S, Dong W, Lan B, Zhao H, Hou L, Zhu L, Xu Y, Lu Y (2019) Flexible carbonic pen ink/carbon fiber paper composites for multifunctional switch-type sensors. Compos Part A Appl Sci Manuf 124:105452. https://doi.org/10.1016/j.compositesa.2019.05.020
33. Araújo DAG, Camargo JR, Pradela-Filho LA, Lima AP, Muñoz RAA, Takeuchi RM, Janegitz BC, Santos AL (2020) A lab-made screen-printed electrode as a platform to study the effect of the size and functionalization of carbon nanotubes on the voltammetric determination of caffeic acid. Microchem J 158:105297. https://doi.org/10.1016/j.microc.2020.105297
34. Liu X, Yao Y, Ying Y, Ping J (2019) Recent advances in nanomaterial-enabled screen-printed electrochemical sensors for heavy metal detection. TrAC Trends Anal Chem 115:187–202. https://doi.org/10.1016/j.trac.2019.03.021
35. Trojanowicz M (2016) Impact of nanotechnology on design of advanced screen-printed electrodes for different analytical applications. TrAC Trends Anal Chem 84:22–47. https://doi.org/10.1016/j.trac.2016.03.027
36. Smith DA, Simpson K, Lo Cicero M, Newbury LJ, Nicholas P, Fraser DJ, Caiger N, Redman JE, Bowen T (2021) Detection of urinary microRNA biomarkers using diazo sulfonamide-modified screen printed carbon electrodes. RSC Adv 11:18832–18839. https://doi.org/10.1039/D0RA09874D
37. Chanarsa S, Jakmunee J, Ounnunkad K (2021) A bifunctional nanosilver-reduced graphene oxide nanocomposite for label-free electrochemical immunosensing. Front Chem 9:245. https://www.frontiersin.org/article/10.3389/fchem.2021.631571
38. Proença CA, Baldo TA, Freitas TA, Materón EM, Wong A, Durán AA, Melendez ME, Zambrano G, Faria RC (2019) Novel enzyme-free immunomagnetic microfluidic device based on Co0.25Zn0.75Fe2O4 for cancer biomarker detection. Anal Chim Acta 1071:59–69. https://doi.org/10.1016/j.aca.2019.04.047

39. de Oliveira TR, Erbereli CR, Manzine PR, Magalhães TNC, Balthazar MLF, Cominetti MR, Faria RC (2020) Early diagnosis of Alzheimer's disease in blood using a disposable electrochemical microfluidic platform. ACS Sensors 5:1010–1019. https://doi.org/10.1021/acssensors.9b02463
40. Fava EL, Silva TA, Prado TMD, Moraes FCD, Faria RC, Fatibello-Filho O (2019) Electrochemical paper-based microfluidic device for high throughput multiplexed analysis. Talanta 203:280. https://doi.org/10.1016/j.talanta.2019.05.081
41. Fava EL, Martimiano do Prado T, Almeida Silva T, Cruz de Moraes F, Censi Faria R, Fatibello-Filho O (2020) New disposable electrochemical paper-based microfluidic device with multiplexed electrodes for biomarkers determination in Urine sample. Electroanalysis 32:1075–1083. https://doi.org/10.1002/elan.201900641
42. Vinoth R, Nakagawa T, Mathiyarasu J, Mohan AMV (2021) Fully printed wearable microfluidic devices for high-throughput sweat sampling and multiplexed electrochemical analysis. ACS Sensors 6:1174–1186. https://doi.org/10.1021/acssensors.0c02446
43. Sethi J, Suhail A, Safarzadeh M, Sattar A, Wei Y, Pan G (2021) NH2 linker for femtomolar label-free detection with reduced graphene oxide screen-printed electrodes. Carbon N Y 179:514–522. https://doi.org/10.1016/j.carbon.2021.04.074
44. Valverde A, Montero-Calle A, Barderas R, Calero M, Yáñez-Sedeño P, Campuzano S, Pingarrón JM (2021) Electrochemical immunoplatform to unravel neurodegeneration and Alzheimer's disease through the determination of neurofilament light protein. Electrochim Acta 371:137815. https://doi.org/10.1016/j.electacta.2021.137815
45. Bagheri Hashkavayi A, Cha BS, Hwang SH, Kim J, Park KS (2021) Highly sensitive electrochemical detection of circulating EpCAM-positive tumor cells using a dual signal amplification strategy. Sensors Actuators B Chem 343:130087. https://doi.org/10.1016/j.snb.2021.130087
46. Rashid JIA, Kannan V, Ahmad MH, Mon AA, Taufik S, Miskon A, Ong KK, Yusof NA (2021) An electrochemical sensor based on gold nanoparticles-functionalized reduced graphene oxide screen printed electrode for the detection of pyocyanin biomarker in *Pseudomonas aeruginosa* infection. Mater Sci Eng C 120:111625. https://doi.org/10.1016/j.msec.2020.111625
47. Rauf S, Lahcen AA, Aljedaibi A, Beduk T, de Oliveira Filho JI, Salama KN (2021) Gold nanostructured laser-scribed graphene: A new electrochemical biosensing platform for potential point-of-care testing of disease biomarkers. Biosens Bioelectron 180:113116. https://doi.org/10.1016/j.bios.2021.113116
48. Sung R, Heo YS (2021) Sandwich ELISA-based electrochemical biosensor for Leptin in control and diet-induced obesity mouse model. Biosens 11:7. https://doi.org/10.3390/bios11010007
49. Zhang B, Liang Y, Bo L, Chen M, Huang B, Cao Q, Wei J, Li T, Cai X, Ye X (2021) A PCR-free screen-printed magnetic electrode for the detection of circular RNA from hepatocellular cancer based on a back-splice junction. RSC Adv 11:17769–17774. https://doi.org/10.1039/D1RA01033F
50. Muñoz-San Martín C, Gamella M, Pedrero M, Montero-Calle A, Pérez-Ginés V, Camps J, Arenas M, Barderas R, Pingarrón JM, Campuzano S (2021) Anticipating metastasis through electrochemical immunosensing of tumor hypoxia biomarkers. Anal Bioanal Chem 414:399. https://doi.org/10.1007/s00216-021-03240-8
51. Chen M, Wu D, Tu S, Yang C, Chen D, Xu Y (2021) A novel biosensor for the ultrasensitive detection of the lncRNA biomarker MALAT1 in non-small cell lung cancer. Sci Rep 11:3666. https://doi.org/10.1038/s41598-021-83244-7
52. Leung W-H, Pang C-C, Pang S-N, Weng S-X, Lin Y-L, Chiou Y-E, Pang S-T, Weng W-H (2021) High-sensitivity dual-probe detection of urinary miR-141 in cancer patients via a modified screen-printed carbon electrode-based electrochemical biosensor. Sensors 21:3183. https://doi.org/10.3390/s21093183
53. Chen M, Li H, Su X, Wu R, Feng H, Shi X, Liang J, Chen J, Li G (2021) Label-free electrochemical aptasensor based on reduced graphene oxide–hemin–chitosan nanocomposite for the determination of glypican-3. New J Chem 45:8608–8618. https://doi.org/10.1039/D1NJ00633A

54. Pothipor C, Jakmunee J, Bamrungsap S, Ounnunkad K (2021) An electrochemical biosensor for simultaneous detection of breast cancer clinically related microRNAs based on a gold nanoparticles/graphene quantum dots/graphene oxide film. Analyst 146:4000–4009. https://doi.org/10.1039/D1AN00436K
55. Ma F, Ge G, Fang Y, Ni E, Su Y, Cai F, Xie H (2021) Prussian blue-doped PAMAM dendrimer nanospheres for electrochemical immunoassay of human plasma cardiac troponin I without enzymatic amplification. New J Chem 45:9621–9628. https://doi.org/10.1039/D1NJ01506K
56. Materon EM, Wong A, Gomes LM, Ibáñez-Redín G, Joshi N, Oliveira ON, Faria RC (2021) Combining 3D printing and screen-printing in miniaturized, disposable sensors with carbon paste electrodes. J Mater Chem C 9:5633–5642. https://doi.org/10.1039/D1TC01557E
57. Kalasin S, Sangnuang P, Khownarumit P, Tang IM, Surareungchai W (2020) Salivary creatinine detection using a Cu(I)/Cu(II) catalyst layer of a supercapacitive hybrid sensor: a wireless IoT device to monitor kidney diseases for remote medical mobility. ACS Biomater Sci Eng 6:5895–5910. https://doi.org/10.1021/acsbiomaterials.0c00864
58. Amor-Gutiérrez O, Costa-Rama E, Arce-Varas N, Martínez-Rodríguez C, Novelli A, Fernández-Sánchez MT, Costa-García A (2020) Competitive electrochemical immunosensor for the detection of unfolded p53 protein in blood as biomarker for Alzheimer's disease. Anal Chim Acta 1093:28–34. https://doi.org/10.1016/j.aca.2019.09.042
59. Silva RR, Raymundo-Pereira PA, Campos AM, Wilson D, Otoni CG, Barud HS, Costa CAR, Domeneguetti RR, Balogh DT, Ribeiro SJL, Oliveira ON Jr (2020) Microbial nanocellulose adherent to human skin used in electrochemical sensors to detect metal ions and biomarkers in sweat. Talanta 218:121153. https://doi.org/10.1016/j.talanta.2020.121153
60. Neves MMPS, Nouws HPA, Santos-Silva A, Delerue-Matos C (2020) Neutrophil gelatinase-associated lipocalin detection using a sensitive electrochemical immunosensing approach. Sensors Actuators B Chem. 304:127285. https://doi.org/10.1016/j.snb.2019.127285
61. Freitas M, Nouws HPA, Keating E, Delerue-Matos C (2020) High-performance electrochemical immunomagnetic assay for breast cancer analysis. Sensors Actuators B Chem. 308:127667. https://doi.org/10.1016/j.snb.2020.127667
62. Shin Low S, Pan Y, Ji D, Li Y, Lu Y, He Y, Chen Q, Liu Q (2020) Smartphone-based portable electrochemical biosensing system for detection of circulating microRNA-21 in saliva as a proof-of-concept. Sensors Actuators B Chem 308:127718. https://doi.org/10.1016/j.snb.2020.127718
63. Valverde A, Serafín V, Garoz J, Montero-Calle A, González-Cortés A, Arenas M, Camps J, Barderas R, Yáñez-Sedeño P, Campuzano S, Pingarrón JM (2020) Electrochemical immunoplatform to improve the reliability of breast cancer diagnosis through the simultaneous determination of RANKL and TNF in serum. Sensors Actuators B Chem 314:128096. https://doi.org/10.1016/j.snb.2020.128096
64. Razzino CA, Serafín V, Gamella M, Pedrero M, Montero-Calle A, Barderas R, Calero M, Lobo AO, Yáñez-Sedeño P, Campuzano S, Pingarrón JM (2020) An electrochemical immunosensor using gold nanoparticles-PAMAM-nanostructured screen-printed carbon electrodes for tau protein determination in plasma and brain tissues from Alzheimer patients. Biosens Bioelectron 163:112238. https://doi.org/10.1016/j.bios.2020.112238
65. Soda N, Umer M, Kashaninejad N, Kasetsirikul S, Kline R, Salomon C, Nguyen N-T, Shiddiky MJA (2020) PCR-free detection of long non-coding HOTAIR RNA in Ovarian cancer cell lines and plasma samples. Cancers 12:2233. https://doi.org/10.3390/cancers12082233
66. Piguillem SV, Gamella M, García de Frutos P, Batlle M, Yáñez-Sedeño P, Messina GA, Fernández-Baldo MA, Campuzano S, Pedrero M, Pingarrón JM (2020) Easily multiplexable immunoplatform to assist heart failure diagnosis through amperometric determination of galectin-3. Electroanalysis 32:2775–2785. https://doi.org/10.1002/elan.202060323
67. Gui J-C, Han L, Du C-X, Yu X-N, Hu K, Li L-H (2021) An efficient label-free immunosensor based on ce-MoS2/AgNR composites and screen-printed electrodes for PSA detection. J Solid State Electrochem 25:973–982. https://doi.org/10.1007/s10008-020-04872-z
68. Ibáñez-Redín G, Joshi N, do Nascimento GF, Wilson D, Melendez ME, Carvalho AL, Reis RM, Gonçalves D, Oliveira ON (2020) Determination of p53 biomarker using an electrochemical

immunoassay based on layer-by-layer films with NiFe2O4 nanoparticles. Microchim Acta 187:619. https://doi.org/10.1007/s00604-020-04594-z
69. Ibáñez-Redín G, Materon EM, Furuta RHM, Wilson D, do Nascimento GF, Melendez ME, Carvalho AL, Reis RM, Oliveira ON, Gonçalves D (2020) Screen-printed electrodes modified with carbon black and polyelectrolyte films for determination of cancer marker carbohydrate antigen 19-9. Microchim Acta 187:417. https://doi.org/10.1007/s00604-020-04404-6
70. Maghsoudi AS, Hassani S, Akmal MR, Ganjali MR, Mirnia K, Norouzi P, Abdollahi M (2020) An electrochemical aptasensor platform based on flower-like gold microstructure-modified screen-printed carbon electrode for detection of serpin A12 as a Type 2 DIABETES BIOMArker. Int J Nanomed 15:2219
71. Povedano E, Ruiz-Valdepeñas Montiel V, Gamella M, Serafín V, Pedrero M, Moranova L, Bartosik M, Montoya JJ, Yáñez-Sedeño P, Campuzano S, Pingarrón JM (2020) A novel zinc finger protein–based amperometric biosensor for miRNA determination. Anal Bioanal Chem 412:5031–5041. https://doi.org/10.1007/s00216-019-02219-w
72. Sethi J, Van Bulck M, Suhail A, Safarzadeh M, Perez-Castillo A, Pan G (2020) A label-free biosensor based on graphene and reduced graphene oxide dual-layer for electrochemical determination of beta-amyloid biomarkers. Microchim Acta 187:288. https://doi.org/10.1007/s00604-020-04267-x
73. Singh V (2019) Quantum dot decorated multi-walled carbon nanotube modified electrochemical sensor array for single drop insulin detection. Mater Lett 254:415–418. https://doi.org/10.1016/j.matlet.2019.07.122
74. Karami P, Bagheri H, Johari-Ahar M, Khoshsafar H, Arduini F, Afkhami A (2019) Dual-modality impedimetric immunosensor for early detection of prostate-specific antigen and myoglobin markers based on antibody-molecularly imprinted polymer. Talanta 202:111–122. https://doi.org/10.1016/j.talanta.2019.04.061
75. Si Y, Park JW, Jung S, Hwang G-S, Park YE, Lee JE, Lee HJ (2019) Voltammetric layer-by-layer biosensor featuring purine nucleoside phosphorylase and chitosan for inosine in human serum solutions. Sensors Actuators B Chem. 298:126840. https://doi.org/10.1016/j.snb.2019.126840
76. Ibáñez-Redín G, Furuta RHM, Wilson D, Shimizu FM, Materon EM, Arantes LMRB, Melendez ME, Carvalho AL, Reis RM, Chaur MN, Gonçalves D, Oliveira ON Jr (2019) Screen-printed interdigitated electrodes modified with nanostructured carbon nano-onion films for detecting the cancer biomarker CA19-9. Mater Sci Eng C 99:1502–1508. https://doi.org/10.1016/j.msec.2019.02.065
77. Wang Y, Luo J, Liu J, Sun S, Xiong Y, Ma Y, Yan S, Yang Y, Yin H, Cai X (2019) Label-free microfluidic paper-based electrochemical aptasensor for ultrasensitive and simultaneous multiplexed detection of cancer biomarkers. Biosens Bioelectron 136:84–90. https://doi.org/10.1016/j.bios.2019.04.032
78. Guerrero S, Cadano D, Agüí L, Barderas R, Campuzano S, Yáñez-Sedeño P, Pingarrón JM (2019) Click chemistry-assisted antibodies immobilization for immunosensing of CXCL7 chemokine in serum. J Electroanal Chem 837:246–253. https://doi.org/10.1016/j.jelechem.2019.02.043
79. Valverde A, ben Hassine A, Serafín V, Muñoz-San Martín C, Pedrero M, Garranzo-Asensio M, Gamella M, Raouafi N, Barderas R, Yáñez-Sedeño P, Campuzano S, Pingarrón JM (2020) Dual amperometric immunosensor for improving cancer metastasis detection by the simultaneous determination of extracellular and soluble circulating fraction of emerging metastatic biomarkers. Electroanalysis 32:706–714. https://doi.org/10.1002/elan.201900506
80. Putnin T, Ngamaroonchote A, Wiriyakun N, Ounnunkad K, Laocharoensuk R (2019) Dually functional polyethylenimine-coated gold nanoparticles: a versatile material for electrode modification and highly sensitive simultaneous determination of four tumor markers. Microchim Acta 186:305. https://doi.org/10.1007/s00604-019-3370-4
81. Chen S, Wang Z, Cui X, Jiang L, Zhi Y, Ding X, Nie Z, Zhou P, Cui D (2019) Microfluidic device directly fabricated on screen-printed electrodes for ultrasensitive electrochemical sensing of PSA. Nanoscale Res Lett 14:71. https://doi.org/10.1186/s11671-019-2857-6

82. Varodi C, Pogacean F, Coros M, Rosu M-C, Stefan-van Staden R-I, Gal E, Tudoran L-B, Pruneanu S, Mirel S (2019) Detection of 8-hydroxy-2′-deoxyguanosine biomarker with a screen-printed electrode modified with graphene. Sensors 19:4297. https://doi.org/10.3390/s19194297

Hybrids of Conducting Polymers and Carbon-Based Materials Aiming Biosensors Applications

Fábio Ruiz Simões, Gabriela Martins de Araújo, and Milton Alexandre Cardoso

Abstract Hybrids of intrinsic conducting polymers (CPs) and carbon-based materials are frequently studied for technological application in several areas. This chapter discusses recent applications of these hybrids in bioelectrochemistry, with a focus on biosensors. These hybrids presented synergic characteristics such as large surface, significant electrical conductivity, and outstanding thermal and mechanical properties improving the analytical responses as selectivity, sensitivity, reproducibility, and response time, of the biosensors. Thus, these hybrids have been extensively explored in the development of non-enzymatic and enzymatic biosensors and used in medical, environmental, food, and pharmaceutical analysis. Biosensors based on these CPs/carbon-based hybrids are simple, specific, and sensitive as well as enable manufacture disposable, and portable miniature device.

Keywords Hybrids · Biosensors · Bioelectrochemistry · Conducting polymers · Carbon-based materials

1 Introduction

Research around the world has associated different materials to manufacture composites to improve physical or chemical characteristics. Hybrid materials appear in this scenario as composites that have strict control over their components. Some of the most common hybrid materials are developed by associating CPs with carbon-based materials. Variations in the proportion between them, in their form of manufacture and application, may favor specific characteristics [1, 2]. In electrochemical biosensors, these can be expressed through increased conductivity, selectivity, mechanical strength, stability, efficiency, among others.

Present Address:
F. R. Simões (✉)
Federal University of São Paulo, UNIFESP, Institute of Marine Sciences, São Paulo, SP, Brazil
e-mail: fabio.simoes@unifesp.br

G. M. de Araújo · M. A. Cardoso
Graduate Program in Chemistry: Sustainability Science and Technology, UNIFESP, Campus Diadema, Diadema, SP, Brazil

F. N. Crespilho (ed.), *Advances in Bioelectrochemistry Volume 3*,
https://doi.org/10.1007/978-3-030-97921-8_6

Hybrids based on CPs and carbon-based materials, such as graphite [3–5], carbon black [6, 7], grapheme [8–11], and carbon nanotubes [9, 12–16], have been developed for decades in several applications such as reinforce materials [15, 17], sensors [13, 18–21], batteries [8, 22, 23], and solar cells [24, 25]. The synergic effects can be achieved once CPs have mechanical properties of polymers and the conductivity adjusted in a wide range while the carbon-based material can improve the conductivity, show high specific area, and elevated strength modulus [26–29]. The high specific area allied to the conductivity makes these hybrids promising materials in bioelectrochemistry such as sensors and charge storage devices [10, 13, 30–34].

These hybrid materials present characteristics as excellent thermal and mechanical properties, high electrical conductivity, and large electrode surface area. Besides it is possible to immobilize enzymes in these hybrids when applied in biosensor due to the presence of functional groups that allowed the monitoring of drugs in biological fluids [35]. They are also very used in the fabrication of non-enzymatic biosensors like materials for sensing purposes. They present a low-cost, fast, and simple method improving the electrochemical response of the biosensor. Furthermore, the non-enzymatic electrochemical biosensors present great chemical stability because these devices are enzyme-free [20, 36].

Enzymatic and non-enzymatic electrochemical biosensors based on these hybrids have been widely applied in food, medical, environmental, and pharmaceutical areas due to the characteristics of these hybrid materials to enable improve the analytical response as sensitivity, selectivity, well as response time, and reproducibility of these biosensors system [33, 37].

2 Biosensors Based on Hybrids of Conducting Polymers and Carbon-Based Materials

Electrochemical biosensors based on hybrids of CPs and carbon-based materials are applied in several areas as: environmental applications by the detection of pesticides [38], phenolic compounds [39, 40], metals [41], among others; medical by to control of diseases, for example, diabetes through of the monitoring of glucose [36, 42–45], diagnostics and biomedical detection of DNA specific sequence [46, 47], among others. Due to its simplicity, specificity, sensitivity as well as enable to perform repetitive measurements with miniaturized and portable devices and low cost [35].

2.1 Environmental

OPs and carbamates (CBs) pesticides (Fig. 1) play a key role in agricultural productivity, due to their good effectiveness for insect eradication and low price.

Fig. 1 Simplified chemical structure of organophosphate with free binding sites (R′, R″ and R‴) and chemical structure of carbamate with free binding sites (R′ and R″)

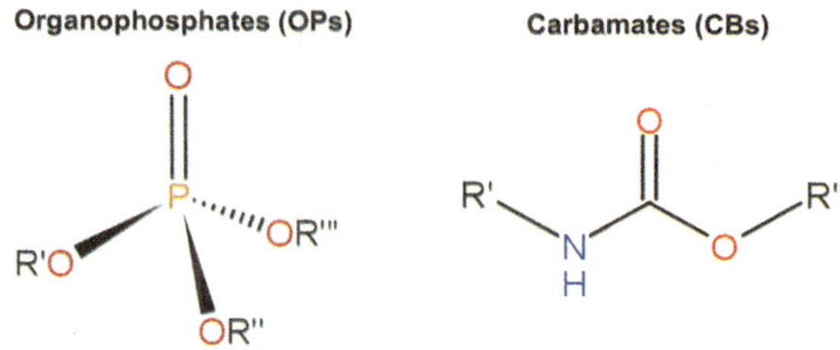

Therefore, early detection of the OPs and CBs can mitigate the dangers of their accumulation for a long period of time in the environment may negatively affect food safety and human health [32, 48, 49]. OPs and CBs directly affect the central nervous system out of living organisms, since they cause inhibition of the AChE, may cause if used in excess memory decline, muscular paralysis, convulsions, respiratory failure, and even death [50, 51]. In addition to the OPs and CBs, also worth mentioning phenolic compounds that are organic compounds that contain aromatic rings with hydroxyl functional groups as shown in Fig. 2.

These compounds originate from industrial products and plant materials or waste, which infiltrate the soil as leachate or as particulate material; therefore, even present at a low concentration, it is harmful to the environment causing disturbances and impacts that unbalance biogeochemical cycles and the marine food web, and also

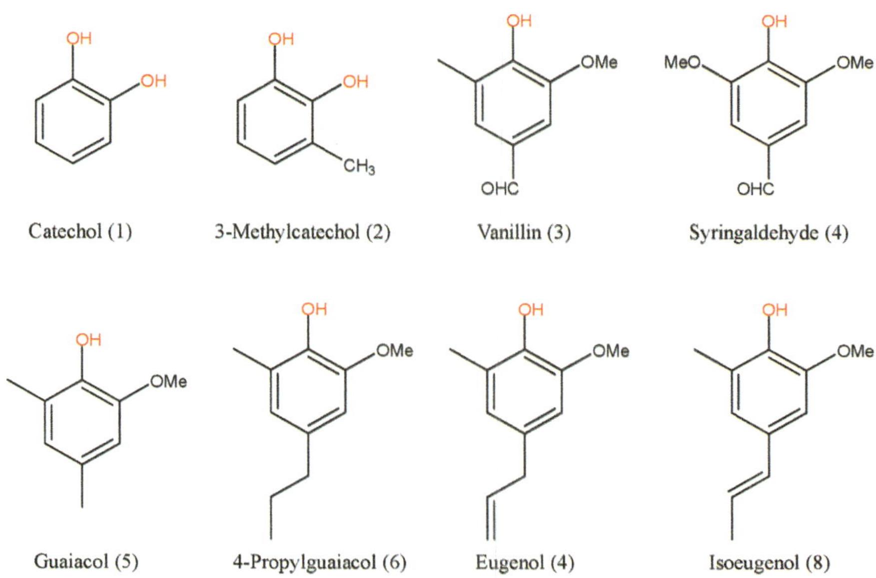

Fig. 2 Chemical structure of some phenolic compounds

causing irreversible toxic effects in several organisms; in addition, they can affect human health [40, 52].

Some heavy metals in small quantities are essentials important for human health such as zinc, manganese, vanadium, iron, copper, cobalt, strontium, and molybdenum. However, in disordered levels, heavy metals like cadmium, mercury, and lead can also cause damage to humans and the environment, as their metal ions are not biodegradable and can accumulate in soft tissue, causing many diseases, in addition to metals like antimony, arsenic, lead, cadmium, mercury, chromium, impact surface water systems [53, 54].

The determination of these compounds has been widely employed using various methods like mass spectrometry (MS), high-performance liquid chromatography (HPLC), and gas chromatography (GC). However, these methods are disadvantageous in relation to sample pretreatment, lengthy turnaround time, expensive equipment requiring extensive labor and analytical resources. The application of these techniques in the determination of analyte (OPs and CBs), phenolic compounds, and heavy metals have limitations such as real-time monitoring and rapid analyses with low costs [40, 49].

Electrochemical biosensors are promising for direct monitoring of OPs [55] and CBs [33] phenolic compounds [40] and heavy metals [41] able to detect its residues at low concentrations. Direct monitoring through biosensors allows for fast responses, high sensitivity, with great versatility, low limit of quantification (LoQ) and low limit of detection (LoD), and results that can be reproduced and compared with classical analytical techniques, such as MS and chromatography [40, 49, 55]. Biosensors based on hybrids of carbon-based materials and CPs are used to the monitoring as well detecting of pesticides, phenolic compounds, and heavy metals to overcome the disadvantages of other analytical techniques due to combining the properties of both the materials classes [33, 40, 41, 55].

Some works describe the development of non-enzymatic biosensors for phenolic compounds detection using hybrid materials using stable, simple, low-cost, and effective method as well as report enzymatic biosensors for pesticides related to the inhibition of the activity of AChE. Besides that, the enzyme is very applied in the fabrication of biosensors for pesticide determination because this enzyme operates as a powerful recognition element.

To manufacture the biosensor, it is fundamental to immobilize the enzyme on the electrode. The effectiveness of the device is directly related to this step. The interaction between enzyme and the working electrode directly influences the catalytically achievement of the electrode. In the enzymatic biosensors used for the pesticide detection, CBs, and OPs act to inhibit the enzyme reaction AChE. The inhibition is reversible on a small scale in the CBs, while most OPs pesticides prompt an irreversible inhibition in the absence of the analyte in the supporting electrolyte. Due to the applied potential, the substrate ATCh is converted into acetic acid and thiocholine when the pesticide is absent in the solution. However, in the presence of the pesticide that acts as an inhibitor, the transformation of ATCh is reduced or null. Additionally, the concentration of analyte such as CBs and OPs in the sample generate an inversely proportional response in the oxidation current signal [56].

Kesik et al. [57] amperometric AChE biosensor based on poly(4-(2,5-di(thiophen-2-yl)-1H-pyrrol-1-yl)benzenamine)(poly(SNS-NH_2)) and MWCNT modified electrode was developed for the detection of OPs pesticides. The functionalized MWCNT was used to modify the graphite electrode, and then the poly (SNS-NH2) was synthesized electrochemically onto a modified electrode with MWCNT. The enzyme immobilization occurred by covalent linkage on the surface of a modified graphite electrode with MWCNT/poly (SNS-NH_2). This procedure is applying a simultaneous coupling method two-step of carbodiimide, based on the formation of bonds of the free amino groups of poly (SNS-NH_2) and free carboxylic acid groups of f-MWCNT with the AChE. CV, EIS, SEM, and X-ray photoelectron spectroscopy (XPS) were applied to observe morphological changes and electrochemical characterizations of the electrode. After the enzyme immobilization on the electrode surface, a morphological change was observed through the SEM images. XPS technique showed through specific peaks that the functionalization and immobilization processes were realized with success. The Nyquist diagrams from EIS results showed the immobilization of enzyme was obtained onto the graphite electrode with MWCNT/poly (SNS-NH_2) due to a large increase in the diameter of the semicircle, showing that the formation of the layer hindered the flow of the electrons. CV results confirmed the fixation of the biomolecule on electrode surface due to show a decrease of oxidation current (245 μA). The developed biosensor exhibited a linear range of 0.05 mM and 8.00 mM, a LoD of 0.09 mM, sensitivity of 24.16 $\mu A\ mM^{-1}\ cm^{-2}$ and a fast amperometric response time (6 s), for acetylthiocholine chloride as the substrate. The inhibition responses of the parathion, chlorfenvinphos, and paraoxon were used in the biosensor in fortified tap water samples on the activity of the AChE. So, the authors concluded that the developed biosensor showed similar results to those obtained for chromatographic techniques for the determination of parathion, chlorfenvinphos and paraoxon.

Zheng et al. [40] produced electrochemistry biosensor free of enzyme based on polyphenol oxidase (PPO) into composites of PAni, graphene polyacrylonitrile (PAN)-nanostructured, for the detection of phenolic pollutants in water. A bare platinum electrode was modified with PAni-PAN-graphene from the hybrid assembling method, and then the PPO with acidic chitosan (CS) was immobilized by coating and phase inversion process onto the modified electrode with hybrid. Finally, occurred the cross-linking of the enzyme molecules through glutaraldehyde vapor vessel in contact with the modified electrode. Amperometry responses showed response currents more stable from the modified electrode with the PAni-PAN-graphene hybrid due to excellent conductivity and good biocompatibility. Besides that, the synergistic effect of PAni-PAN-graphene hybrid. The images of SEM showed the morphologies of the materials as well as a big quantity of PPO incorporated inside the PAni-PAN-graphene nanohybrid. The p-cresol was detected utilizing the electrochemistry techniques of EIS and CV as well the amperometry. The CV results indicated that the reaction catalyzed by p-cresol PPO occurred in the biosensor due to showing an increased oxidation current only the PAni-PAN-graphene/CS-PPO in the presence of p-cresol in comparison with the others modified electrodes, since that the direct electrochemistry of PPO. The Nyquist diagrams from EIS result in pH 6.0 PBS showed that the modified electrode based on PAni-PAN-graphene/CS-PPO had the value of charge transfer resistance (R_{ct}) founded increased in small increments comparing with electrode without the PPO (PAni-PAN-graphene/CS), due to non-conductive property of the PPO, hindering the electron transport. Biosensor performance was evaluated with successive addition of p-cresol using amperometry technique. The biosensor showed a LoD of 2.65×10^{-7} M, response time about 5 s and sensitivity of 6.46 μA μM^{-1} cm^{-2}. The biosensor presented high selectivity for the p-cresol, with low inorganic interference such as Zn^{2+}, NO^{3-}, Mg^{2-}, Ca^{2+}, Na^{+}, Co^{2+}, Cl^{-}, $SO4^{2-}$, Fe^{2+}, Fe^{3+}, among other besides organics such the glucose, aniline, and ascorbic acid. Due to the synergistic effect of PAni-PAN-graphene hybrid, specific catalysis of the PPO and good isolation of the graphene. The synergistic effect of hybrid improved the electron transfer ability, anti-interference ability, and amplified the signal of the nanocomposites. The biosensor developed for determination of p-cresol in water samples obtained appropriate results in comparison with the ultra-performance liquid chromatographic (UPLC) method.

Nowadays, electrochemistry biosensors fabricated with carbon-based materials and CPs have stood out in the determination of heavy metals. Some works in the literature describe the determination of toxic heavy metal traces in diversified biological materials. Whether in clinical medicine, natural waters such as ocean, sea, rivers, and wastewater using electrochemical biosensors modified with these hybrids [41]. These biosensors are used to monitor electroactive species (heavy metals) arising from the metabolism of biological components such as enzymes or DNAzymes and cells [41].

As, for example, Yang et al. [41] developed a biosensor to detect mercury ions (Hg^{2+}) through a modified Au electrode from sensitive layer of the nanocomposite based on rGO and PAni immobilized DNA strands due to high-intensity amino groups on the PAni surface. The hybrid material (rGO-PAni) was synthetized by in situ polymerization of aniline using ammonium persulfate. rGO-PAni hybrid was deposited by drop-casting onto Au electrode, and then with the Au/rGO-PAni was dipped in the PBS solution with the DNA. The Au electrode with the composite exhibited a good affinity with the immobilization of T-rich DNA strands, which chose to attach with Hg^{2+} to establish T–Hg^{2+}–T coordination. The Au/rGO-PAni immobilized with DNA formed double helix DNA structure by specific T–Hg^{2+}–T complexes after to be incubated in PBS with different concentrations of Hg^{2+} (Fig. 3I). Field emission scanning electron microscopy (FESEM) and XPS were used to characterize the morphology of the nanocomposite (rGO-PAni). FESEM images showed PAni particles completely covering the rGO layer, obtained the rGO-PAni nanocomposite successfully (Fig. 3II). The XPS results demonstrated characteristic peaks, indicating the fabrication of the rGO-PAni was successful. The electrode Au modified with rGO-PAni and immobilized with DNA was characterized by CV and EIS. The Nyquist diagrams obtained by EIS indicated from the increased of the diameter of the semicircle part after that the Au electrode surface was modified with the rGO-PAni an increase in the R_{ct}. After the DNA immobilization, there was again an increase in the R_{ct} because the DNA increases the impact of the layer affecting the electronic transfer. There was an increase in the film thickness, after Hg^{2+} ions to be added and

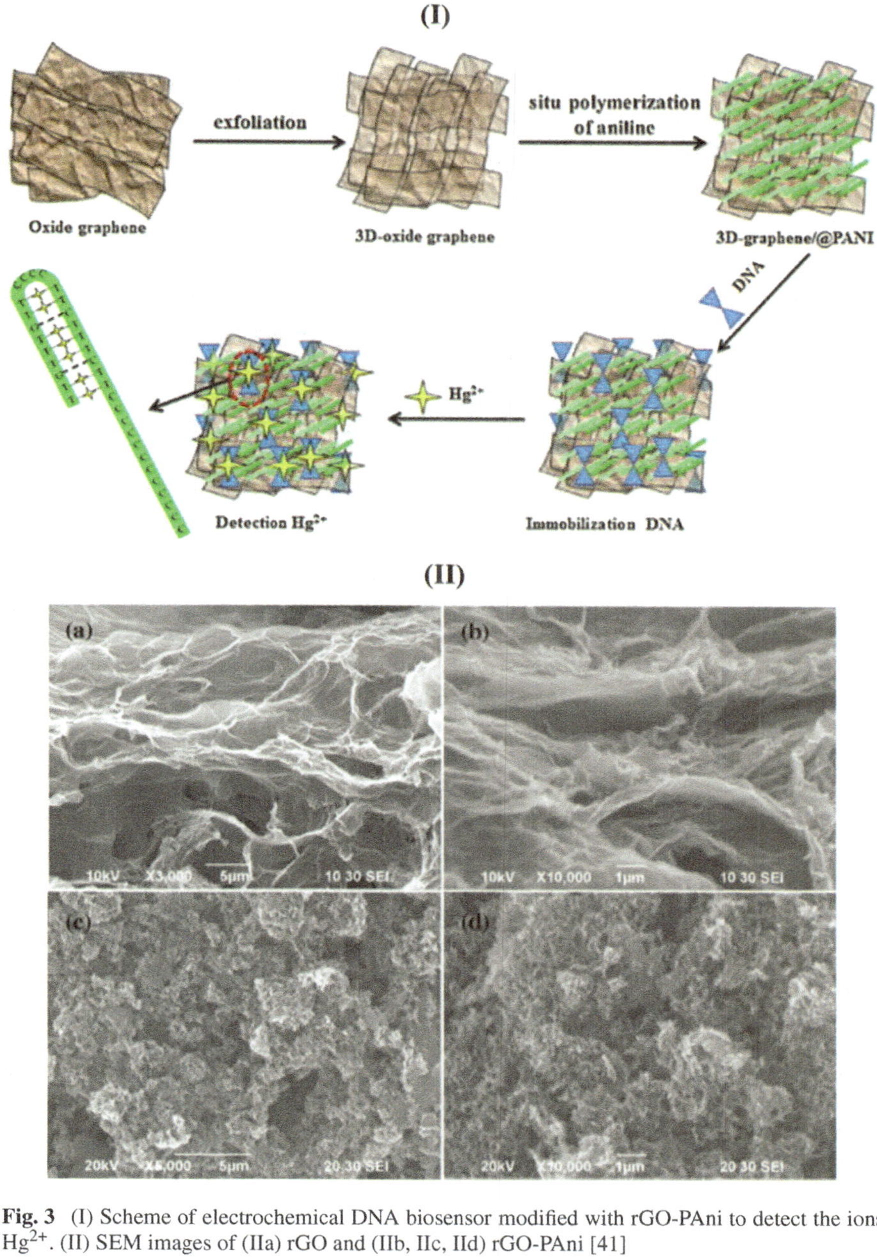

Fig. 3 (I) Scheme of electrochemical DNA biosensor modified with rGO-PAni to detect the ions Hg^{2+}. (II) SEM images of (IIa) rGO and (IIb, IIc, IId) rGO-PAni [41]

the T–Hg2 + –T bond to be formed, resulting in an increased in the R_{ct} value because occurred the formation of an intramolecular T–Hg^{2+}–T duplex. The applicability of the electrochemistry biosensor for the determination of Hg^{2+} was tested using the EIS. The results showed that the method showed high sensitivity due to the R_{ct} increase gradually according to the increase of the mercury ion concentration. The biosensor fabricated presented selectivity and high sensitivity for Hg^{2+} ions in a range from 0.1 to 100 nM with a low LoD of 0.035 nM. The recovery tests for the developed biosensor showed recovery values ranging from 101.2 to 103.4%, suggesting the biosensor as an electrochemical device applicable for determination of Hg^{2+} in real water. The biosensor also showed very good selective response to Hg^{2+} against other metal ions. The authors concluded that the electrochemical biosensor proposed is a good tool for the detection of heavy metal ions with selective and high sensitive in several environmental samples.

2.2 Food

Food safety is of extreme concern for human health since chemical contaminants like antibiotics, heavy metals, pesticides, and pathogenic microorganisms present in food and food products can cause various diseases [58]. Therefore, the role important for consumers and food industries is the quantification and the monitoring of contaminants and food constituents and contaminants. The continuous monitoring for low contaminants levels in food is an area of research that has attracted a lot of attention because the checking food quality is critical for consumer safety and food industry [58, 59]. Traditional analytical techniques such as spectrophotometry, GC, and HPLC coupled with MS used for the analysis of contaminants as the pesticides and metals need expensive instrumentation, are complicated and slow, though these techniques are highly sensitive [60]. Another way for rapid and specific detecting of pesticides these chemical contaminants is electrochemical biosensors because these devices allow the monitoring must often in real time and in situ, providing responses very fast and at small concentration [58]. For example, when foods like apples, cabbages, lettuce are consumed containing malathion and are produced in the body their metabolite the malaoxon which is a hard inhibitor of AChE and more toxic in comparing to malathion [55, 60]. CBs also are based on the inhibiting biocatalytic activity of the AChE impeding through the phosphorylation or carbamylation, the serine residue in the catalytic triad of AChE [58].

Kaur et al. [55] developed an enzymatic bioelectrode based on a nanocomposite of Poly(3,4-ethylenedioxythiophene) (PEDOT) and MWCNT electrochemically deposited on the surface of fluorine doped tin oxide (FTO) for the analysis of malathion OPs in the lettuce samples. The PEDOT-MWCNT on FTO was obtained by CA technique through the electrochemical polymerization. The–COOH functionalization of MWCNT was realized by covalent linkage for the immobilization of the AChE. Characterization studies using Fourier transform infrared spectroscopy (FTIR), FESEM were realized to certify the immobilization of AChE. The FTIR results indicated the immobilization of the enzyme on the surface of the nanocomposites because absorption bands referents to the groups' amine were found in the bioelectrode AChE/PEDOT-MWCNT/FTO. The images obtained by FESEM confirmed the immobilization of AChE onto the nanocomposites due to showing a regular globular structure referent to the AChE. Electrochemical studies were made using CV and EIS techniques. Electroactive surface areas of the electrode were estimated using CV in support electrolyte with KCl and $K_3Fe(CN)_6/K_4Fe(CN)_6$. After immobilization of AChE, the CV showed enzyme the immobilization of enzyme interfered with the electron transport of the PEDOT and MWCNT because the oxidation peak current of enzyme covalently linked presented a decrease. The Nyquist diagrams obtained from the EIS technique showed the R_{ct} value increased after immobilization process of AChE onto the surface electrode modified with PEDOT-MWCNT/FTO for the same reason already mentioned in the CV results. About the catalytic activity of the enzyme on the electrode AChE/PEDOT-MWCNT/FTO was observed by the advance in the oxidation peak in the presence of the acetylthiocholine chloride (ATCl). The bioelectrode inhibition was evaluated using different concentrations of malathion and DPV, and it was observed the reduction of the peak currents with the growth of malathion concentration. The LoD was of 1 fM, in a linear range 1 fM to 1 μM, and the inhibited enzyme was regenerated by 2-pyridine aldoxime methiodide in 99%. The application of the developed method was evaluated in the lettuce sample. The results of the recovery of malathion were between 96 and 98% with the developed biosensor. The authors concluded that this bioelectrode can be a new possibility for the determination of OP.

Cesarino et al. [58] fabricated an electrochemical AChE biosensor based on a modified glassy carbon electrode (GCE) with PAni and MWCNT to detect CBs pesticides in food samples (cabbage, apple, and broccoli). The GCE electrode was dipped in a solution containing aniline and MWCNTCV technique, and the hybrid was electropolymerized and after was deposited AChE on the GCE/PAni/MWCNT by drop-casting. The images obtained by FESEM indicated that the amorphous material linked to the PAni nitrogen groups is related to the enzyme besides to be observed the coverage of the MWCNT with PAni. Square wave voltammetry (SWV) technique was applied in the development and characterization of the MWCNT/PAni/AChE biosensor. SWV results showed that the biosensor with MWCNT increased the current oxidation peak of the dithio-bis-choline in almost three times due to the MWCNT behaving as an electrocatalyst, besides to get better the electrochemical processes of the enzymatic product. The voltammograms obtained by SWV presented peak currents lower, after the additions of carbaryl, despite the same potential obtained for thiocholine oxidation, indicating the inhibition of the activity enzyme with these results, it is possible to conclude that the GCE/MWCNT/PAni/AChE biosensor can determine carbaryl besides of can be restored because CBs inhibition to be reversible. The electrochemical biosensor GCE/MWCNT/PAni/AChE was evaluated in the detection of pesticides (methomyl and carbaryl) using CA in food samples. The biosensor showed a LoD of 1.4 and 0.95 $\mu mol\ L^{-1}$, for carbaryl and methomyl. The results of the recovery tests for the methomyl were between 94.6 and 101.2% to cabbage and 95.4 and 101.9% to broccoli. For the carbaryl in the apple samples, the recovery tests were obtained between 97.3 and 103.4%. The results showed the proposed methodology using the biosensor fabricated to determine methomyl in cabbage and broccoli and carbaryl in apple in this paper was next to the obtained using the HPLC procedure. The authors concluded from the results obtained that the amperometric biosensor developed is promising to determine pesticides in food.

Besides evaluating food samples with contaminants such as pesticides, pathogenic microbes among others, some biosensors are also used to verify the nutritional value, quality, and origin of a food product. Quantification of l-amino acids is essentials in the food industry, since elevated l-amino acids cause clinical disorders in human metabolism. In vinegars, beers, and wines, the chiral evaluation of amino acids is fundamental to determine the quality of the product [61]. For example, Lata et al. [62] developed an amperometric l-amino acid (LAA) biosensor using l-amino acid oxidase immobilized covalently in an Au electrode modified with MWCNT, zinc oxide nanoparticles (ZnONPs) and PAni to detected l-amino acids (LAAs) in juices. The ZnONPs/PAni/MWCNT/Au electrode was fabricated electrochemically using the CV technique. After the l-Amino acid oxidase (LAAO) enzyme was immobilized covalently on the ZnONPs/PAni/MWCNT/Au electrode using carbodiimide (EDC) and N-hydroxysuccinimide (NHS) activated MWCNT (Fig. 4I). The SEM images confirmed the immobilization of LAAO due shows a globular structure morphology referent to the covalent interaction between MWCNT/ZnONPs/PAni/Au electrode with enzyme (Fig. 4II). CV results showed the oxidation and reduction of the immobilized enzyme through the redox peak with oxidative/reductive current for the MWCNT/ZnONPs/PAni/Au. EIS results showed that the presence of MWCNT and ZnONPs inside the PAni matrix made the composite to present more active sites for Faradaic reactions and the electron transport was faster because the Nyquist diagram presented a decrease in the R_{ct} comparing with the Nyquist diagram of the PAni/Au. The interference study using the amperometry technique of the LAA biosensor showed that fructose, ascorbic acid, glucose, uric acid and glycine showed insignificant interference. The LAA biosensor resulted in a linear range of 0.001–70 mM, sensitivity of 46.82 AmM^{-1} cm^{-2} g, and response of 4 s with long stability (120 days at 4 °C). The LAA biosensor was applied for the detection of LAA in juices using amperometry. The average recovery of l-Amino acid (l-phenylalanine) was 96.6 and 95.7% to fruit juice with 5 and 10 mM indicated good reliability for the biosensor fabricated.

The human needs several vitamin components that are essential to keep the individual healthy. However, many of these substances cannot be synthesized by humans and need to be ingested through food. Ascorbic acid (vitamin C) is an example of an important vitamin for human health. It influences collagen production, immune response, osteogenesis, healing processes, neurotransmitter biosynthesis, among others. The lack of ascorbic acid in the body can cause infertility, reduced immunity, and anemia, among other pathologies. Thus, the determination of this analyte in food samples helps to mitigate problems related to deficiency of this vitamin in the human body [62].

Lu et al. [62] developed a biosensor though the modification of GCE with a graphene–PEDOT film with ascorbate oxidase (AO) entrapped for the detection

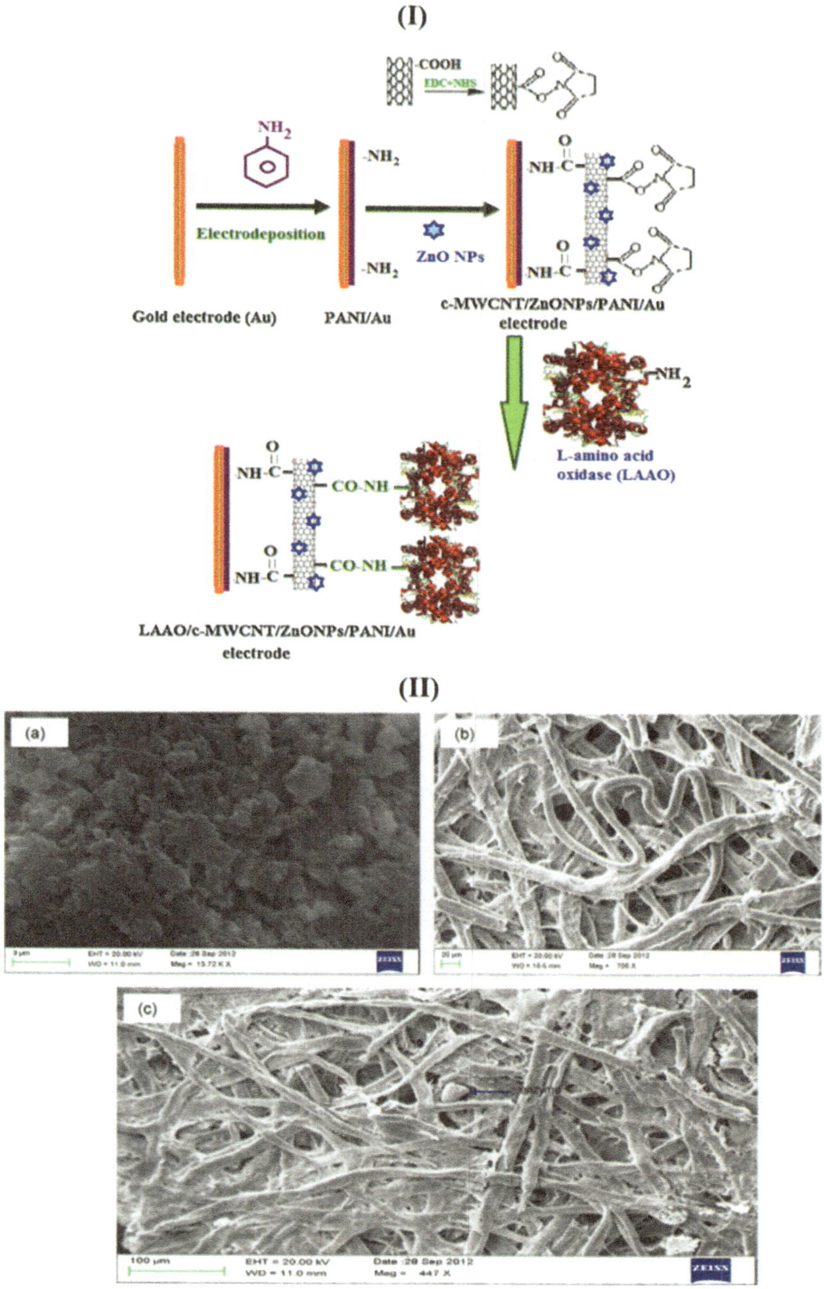

Fig. 4 (I) Schematic illustration of the fabrication of LAAO/MWCNT/ZnONPs/PAni/Au electrode. SEM images (IIa) PAni/Au electrode, (IIb) MWCNT/ZnONPs/PAni/Au and (IIc) LAAO/MWCNT/ZnONPs/PAni/Au electrode [61]

of ascorbic acid in fruit juices samples. By CV, they electrochemically deposited nanocomposite film of graphene–PEDOT and graphene–PEDOT/AO using GCE electrode. To characterize the film was used UV–vis spectroscopy (UV–vis) and SEM. The graphene–PEDOT nanocomposite. SEM images showed a morphology of wrinkled graphene covered by PEDOT providing a good matrix to immobilize biological molecules (Figs. 5IIa–c). UV–vis spectrum of the PEDOT film showed an absorbance band-centered attributed to PEDOT oligomers (Fig. 5IId). Besides that, the UV-vis results showed the formation of graphene–PEDOT nanocomposite, because the absorbance intensity was lower for the nanocomposite film comparing with PEDOT film, besides showing an absorption peak at high wavelength related to bigger thickness of the graphene–PEDOT deposited on ITO and to the π-polaron transition among PEDOT and graphene. CV results showed the graphene–PEDOT nanocomposite film presented increase in the peak current comparing with the polymer film indicating the formation of graphene–PEDOT film. However, the peak current decreased with the AO entrapped onto the graphene–PEDOT film, because the AO block interfacial charge transfer operates as an insulating layer. The biosensor was evaluated with successive additions of ascorbic acid by CA. The results showed the nanocomposite electrode presented a large current in comparison to the PEDOT/AO electrode due to synergism among graphene–PEDOT material and the bigger electroactive properties of the immobilized AO. In order to detect ascorbic acid, the biosensor was tested in fruit juice obtaining 21.89 mg/100 mL in lemon juice (Nongfu brand), 9.91 mg/100 mL in grape juice (Nongfu brand), and 12.87 mg/100 mL in orange juice (Huiyuan brand). The authors concluded that the graphene–PEDOT/AO biosensor presented good sensitivity, good stability, and a wide linear range for ascorbic acid.

2.3 Diseases

In medical applications, the biosensors are growing rapidly, especially the glucose biosensors, non-enzymatic, and enzymatic. Electrochemical glucose biosensors are cheaper, simpler, and faster regarding other methods such as spectroscopy, chromatography, MS, among others, leading to the fast growth of this market valued at $12.8 billion in 2019 [31]. For example, diagnosis of diabetes mellitus can be done through biosensors, requiring a good control over blood glucose. It is worth mentioning that this disease causes much death around the world [63]. There are two types of diabetes mellitus. The type 1 known as diabetes T1DM occurs by the fabrication of insulin through the pancreas. Type 2 is caused due to the body does not utilize all the insulin fabricated and is known as diabetes T2DM [64]. Besides in the production of food and beverage, the estimation of glucose concentration is essential, in which its concentration can vary significantly [31].

The first device that integrated enzyme and GOx was produced in the 60 years by Clark and Lyons [65]. The enzymatic biosensors are simple, efficient, sensitive, and selective and use enzymes to detect glucose indirectly, like GOx. Although due

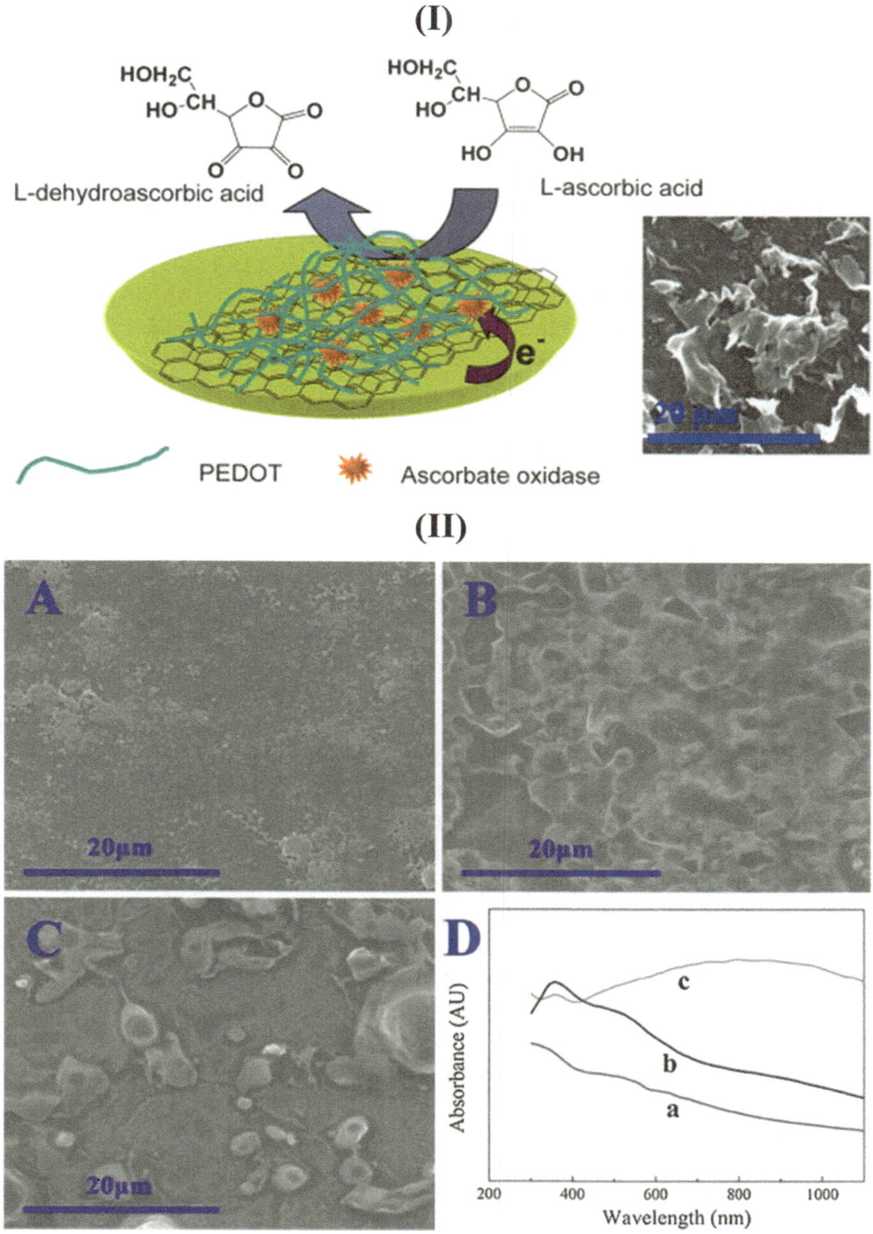

Fig. 5 (I) Scheme of graphene-PEDOT/AO/GCE bionanocomposites for electroanalytical detection of ascorbic acid; (IIa) Images WITHOUT PEDOT film, (IIb) graphene and (IIc) graphene-PEDOT film. (IId) UV-vis spectrum of (a) graphene, (b) PEDOT, and (c) graphene film/PEDOT [62]

to the enzymatic immobilization process, this type of biosensor presents challenges such as reproducibility, high cost, and instability [66, 67]. Since then, there has been an upgrading in the fabrication of biosensors able to carry out the monitoring and detecting of biological metabolites such as glucose in real time with low cost and high reliability [65].

Some works related to the development of the enzymatic glucose biosensor from the hybrids based on CPs such as PAni or PPy and carbon nanomaterial such as graphene or carbon nanotubes, mainly due to the well-established synthetic routes and the biocompatibility of these hybrids. For example, Xue et al. [68] fabricated an amperometric glucose biosensor based on gold nanoparticles (AuNPs), rGO, and PPy nanocomposite. The synthesis of PPy/rGO occurred after added the pyrrole dropwise in an acid solution with GO dispersed under vigorous stirring and at high temperature. PPy/rGO nanocomposite was obtained after 24 h under stirring with reflux. Then the nanocomposite was dispersed in water and was added sodium borohydride for the chemical reduction of the GO and so was joined chloroauric acid to the suspension, and finally, the AuNPs/PPy/rGO was dropped onto a GCE. Already, the AuNPs/PPy/rGO/GOx/Chi electrode was developed by drop-casting of Chi aqueous solution containing GOx and AuNPs/PPy/rGO. The SEM and transmission electron microscopy (TEM) were used for morphological identification of AuNPs/PPy/rGO. The SEM and TEM images showed that The PPy/rGO composite was decorated with AuNPs, resulting in better electrocatalytic activity as compared to O_2 reduction using the CV technique. The glucose biosensor was developed through encapsulation of GO via Chi cross-linking. The AuNPs/PPy/rGO/GOx/Chi electrode was tested in the determination of glucose from the CV technique. CV results showed that by increasing the concentration of glucose catalyzed by GOx, O_2 is consumed, and thus, its reduction peak decreases. The developed biosensor presented linearity of 0.2 to1.2 mM with $R = 0.986$ and sensitivity of 123.8 A mM^{-1} cm^{-2}. Due to anchoring ability, electronic transfer, and biocompatibility, the authors concluded that the AuNPs/PPy/rGO can be promising in encapsulating some other biomolecules.

Non-enzymatic biosensors enable a rapid and cost-effective sensing of glucose by oxidation or reduction with base on the direct electrochemistry of glucose. This type of biosensor has some advantages as it is free from the degradation problem since it does not require an enzymatic immobilization procedure and has high stability, simplicity, and reproducibility [64]. Free enzymes glucose biosensors that using hybrids of CPs and carbon-based nanomaterials have been related in the literature to overcome some limitations already discussed of the enzymatic glucose biosensors and to have a more sensitive and selective performance.

Zhuang et al. [36] developed a free enzyme biosensor for glucose determination based on nickel oxide nanoparticles (NiONPs), PAni, and GO composite. The aniline/GO hybrid was drop cast on the GCE, and then the aniline/GO/GCE was electrodeposited in a deoxygenized $NiCl_2$ solution with KCl using CA technique. The GCE modified was characterized by SEM and CV. SEM images showed that the structure of NiONPs/PAni/GO/GCE was more homogeneous and smaller because the final structure of the hybrid is allowed to obtain NiONPs better distributed and also with smaller size. CV results demonstrated that the

NiONPs/PAni/GO/GCE presented a pair of well-defined redox peaks in alkaline solution. Due to the presence of the PAni make the NiONPs/PAni/GO/GCE more active. In CV results, NiONPs/PAni/GO/GCE showed to glucose oxidation a better electrocatalytic activity, due to the lower Ni content. The applicability of the NiONPs/PAni/GO/GCE was evaluated in an alkaline solution containing glucose of varying concentrations through the CV technique. The results suggested a good electrocatalytic activity of the NiONPs/PAni/GO/GCE due to the transformation from NiO to NiOOH which improves by increasing glucose concentration. The sensitivity study was made through the CA technique. The biosensor exhibited a LoD of 0.5 μM, a linearity range of 2 μM–5.560 mM and sensitivity of 376.22 μA mM^{-1} cm^{-2}. The NiONPs/PAni/GO/GCE biosensor exhibited excellent selectivity to detect glucose in the presence of interferents in complex matrices, such as fetal bovine serum from amperometry technique. The authors concluded that the non-enzymatic glucose biosensor would be promising to detect glucose in clinical diagnosis, food analysis with high sensitivity and sensitivity.

The determination of cholesterol (Fig. 6) is a crucial biomarker for several diseases and has a lot of relevance in clinical analysis, due to this important lipid for humans prevent and diagnostic many diseases like cerebral thrombosis, heart disease, arteriosclerosis, and hypertension [69, 70]. Cholesterol and its acid esters are essential elements of the mammalian cell membrane, composing the nerve tissues and brain. These constituents also form other biological materials, like steroid hormones and bile acid [71, 72].

The cholesterol is carried in the blood, while inside lipoproteins are complexes of lipid and protein. The low-density lipoprotein (LDL) is a lipoprotein known as "bad cholesterol". The LDL carries cholesterol from the liver to arteries of the body, which will be collected in artery walls, leading to deposition in plaque in blood vessels. When it is at a high level, the LDL can cause arterial diseases because the arteries can be blocked by blood clots formed due to buildup of plaques. Another lipoprotein

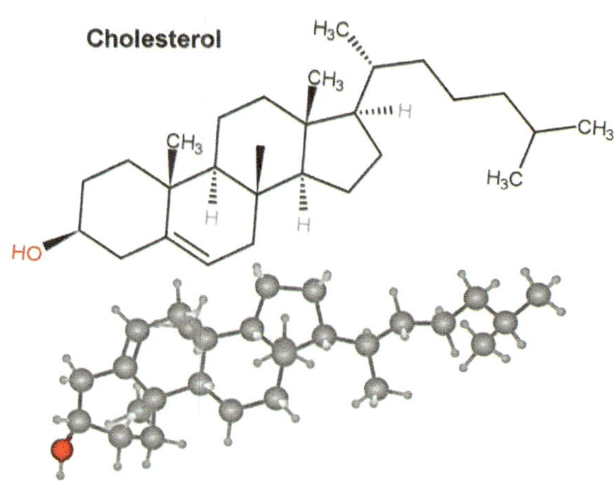

Fig. 6 Chemical structure of cholesterol

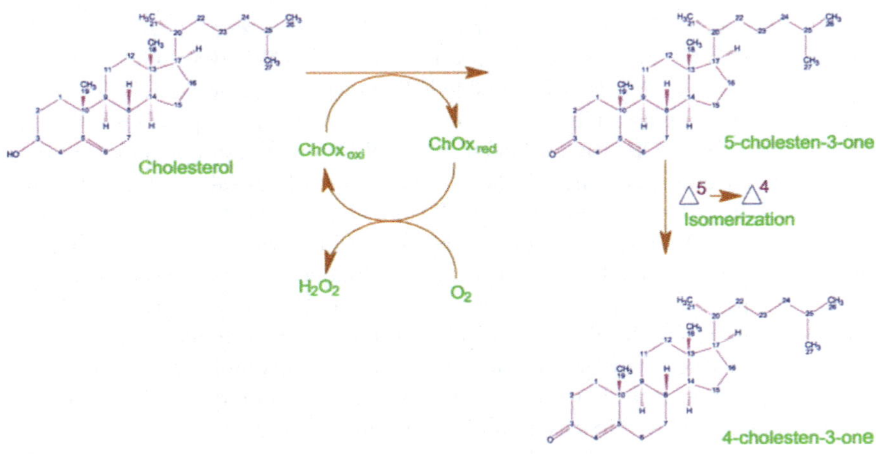

Fig. 7 Cholesterol oxidase enzyme reaction [71]

is the high-density lipoprotein (HDL) known as "good cholesterol" transports the cholesterol of the cells to the liver to then be expelled or broken down from your body. The HDL at a high level can cause heart disease and also stroke [70]. Total blood cholesterol is a measure of LDL cholesterol and triglycerides, and its normal level is around 200 mg/dL in healthy human serum [70, 72]. The fabrication of precision, low-cost, and sensitive biosensors is fundamental to quantitatively determine cholesterol levels in blood with easiness and fastness since the traditional method like spectrophotometry used for cholesterol determination needs difficult sample preparation and expensive instrument [69, 73].

The determination of cholesterol using biosensors can be performed with or without enzymes. Detection methods using enzymatic biosensors based on the immobilization of the cholesterol oxidase (ChOx) on an electrode surface where selective oxidation occurs of cholesterol to cholest-4-en-3-one and hydrogen peroxide as shown in Fig. 7 [70, 72].

1. ChOx corresponds to a Flavin-adenine-dinucleotide (FAD) having flavoenzyme responsible for catalyzing the oxidation of cholesterol to cholestenone. This enzyme is very used in developed of cholesterol enzymatic biosensors [71]. Some works describe the development of the cholesterol electrochemistry biosensor using hybrids of CPs and carbon materials with ChOx attached to the modified electrode. For example, Ruecha et al. [73] developed a paper cholesterol biosensor using a polyvinylpyrrolidone (PVP), graphene, and PAni nanocomposite for the modification via electrospraying. To develop the paper-based biosensor, the modified wax printing method was employed. In this method, the solution of G/PVP/PAni was added in a syringe and then was deposited onto the working electrode (WE) of a paper biosensor via electrospraying. Masks were used to cover the counterelectrode (CE) and reference to avoid their modification. The electrospray system consists of a syringe pump

with stainless syringes and needles, earthing manifold and power supply. After modification, TEM, SEM, and CV techniques were applied to characterize the biosensor. The results presented the dispersion of graphene in the nanocomposite's graphene/PVP/PAni, with structures evenly distributed in the surface area of the modified electrode. The CV results indicated that the modification allowed the amplification of the anodic and cathodic peak current, showing high sensitivity. The fabricated biosensor was used in the detection of cholesterol through CV and CA. The modified graphene electrode/ PVP/PAni detected cholesterol with an amplification of the anodic current signal of approximately 40 times when compared to the unmodified carbon electrode, and this result was obtained through the indirect detection of cholesterol through the quantification of H_2O_2. The modification during the oxidation of H_2O_2 presented excellent electrocatalytic activity due to the conductivity and surface area of the nanostructures. In the amperometric determination, ChOx is coupled to a graphene/PVP/PAni modified electrode obtaining a LoD of 1 μM, a linear range from 50 μM to 10 mM for analyte determination. Interferers were eliminated with anionic surfactant.

2.4 DNA

DNA-based methods like real-time polymerase chain reaction (real-time PCR) are commonly used because they can measure the non-enzymatic biosensors of cholesterol have several advantages such as their low-cost, high stability, easy fabrication, good reproducibility, and free from oxygen limitations. Scientists have developed new biosensors based on the non-enzymatic detection of cholesterol using different electrode materials as well as carbon nanotubes, graphene, Au, Ag, etc., [74]. Nanostructured materials as graphene, carbon nanotubes, and its hybrids with CPs are promising for the achievement of sensitive, stable, and affordable non-enzymatic cholesterol biosensors.

Some works in the literature describe the fabrication of cholesterol non-enzymatic biosensors based on hybrids of conjugated polymers and carbon-based materials. Gautam et al. [75] fabricated a cholesterol non-enzymatic biosensor for cow milk by using a PAni/MWCNT/starch-modified carbon paste electrode (CPE). The composite was made using in situ chemical polymerization of aniline inside suspension of MWCNT and starch. CPE was fabricated adding graphite powder, PAni/MWCNT/starch, and nujol oil. The sensor performance in the presence of cholesterol was studied by the CV, chronocoulometry (CC), CA, and EIS. CV results in the presence of 0.03 mM of cholesterol at different electrodes indicated that the PAni/MWCNT/Starch is a greater electrocatalytic material than PAni/MWCNT, PAni/Starch for cholesterol determination because of synergism of the ternary composite material (based on PAni, MWCNT, and starch) exhibit better electrical and electrochemical properties. Besides that, it was observed a peak oxidation current reversible of cholesterol in the different electrodes in the presence of cholesterol.

CA results showed that the modified electrode surface presented good diffusion rate through the short response time, within 4–6 s, exhibited by biosensor. The CC results showed efficient electrocatalytic oxidation, due to the amount of charge transfer to increase at higher concentrations of cholesterol. The Nyquist diagrams from EIS results showed that with the increasing in the concentration of cholesterol, occurred a decrease in the R_{ct}. In the CV results, during the several cholesterol concentrations 0.032–0.21 mM were observed an anodic wave referent to the oxidation process at 60 mV and a cathodic peak (−15 and −30 mV) associated with the reduction process. Besides that, it was observed that the anodic peak current increased linearly with successive additions of cholesterol in the solution (PBS pH = 5). The sensor showed a linear response in a wide cholesterol concentration range of 0.032–5 mM, a low LoD of 0.01 mM, high sensitivity (800 μAmM^{-1} cm^{-2}). The biosensor was evaluated to quantify cholesterol in cow milk showing linear response and high sensitivity without employing enzyme and additional redox mediator, indicating that the biosensor could be used in the complex matrix. The authors concluded that this PAni/MWCNT/starch composite is promising for the determination of cholesterol using an integrated device with low cost.

The amplification of Human T-lymphotropic Virus-1, nucleic acid sequence-based amplification in mycobacterium tuberculosis (MTB), among others. However, the real-time PCR technique needs expensive equipment and is time-consuming. Besides, this technique isn't employed for the prognosis of disease progression in the early stage of diseases [76–78]. The precise and precocious detection and diagnosis of diseases as cancer are essentials to avoid human death [79] accelerating healing processes or to become better the life of patients. These kinds of techniques are applicable for diagnosing prostate cancer, breast cancer, and various other types of cancer, almost always are applicable when cancer has been spread to other parts of the body.

Electrochemical sensors are advancing in the initial detection of diseases, reducing to mitigate the risk of death. In this context, DNA electrochemical biosensors have been related in the literature in several works aiming for the rapid and sensitive diagnosis for diseases such as cancer through its biomarkers due to their fastness, facility, low cost, and sample preparation easy [80, 81]. Nowadays, electrochemical DNA biosensors have attracted attention in the determination of pathogens, analysis of drugs in biological samples [82], because to show excellent selectivity, high sensitivity, quick response, low cost, and easy functioning.

DNA electrochemical biosensors are mainly based on electrochemical determination of the hybridization of complementary DNA strands on the electrode surface. In the electrochemical DNA hybridization, biosensors occur in a transducer surface to recognize by base-pairing using attachment of a single-stranded (ss) oligonucleotide probe. The modified electrode with the immobilized probe (oligonucleotide) is immersed in a solution containing the complementary sequence of the same oligonucleotide. The hybrid formed from the linking of the two oligonucleotides in electrode surface will result in changes in electrochemical parameters such as conductivity or capacitance [55, 83].

A good DNA probe immobilization technique is essential for the sensing of the DNA biosensor. The most frequently used enzyme immobilization methods employed in the development of electrochemical DNA biosensors are adsorption, covalent bonding, entrapment, and cross-linking [67].

Immobilization by adsorption uses electrostatic interactions, van der Waals forces, and physical interactions to immobilize DNA probes on the working electrode surface. This method is considered inexpensive and simple because do not use extra reagent [67, 84, 85]. Covalent bonding is very used due to this type of bond supplying the strongest attachment among the enzyme and support in comparison with the absorption method. The covalent bonding's form using the side chains in the enzyme with the supports. This technique demonstrates good stability, flexibility, high binding strength and prevents desorption of DNA probe monolayer in the electrode surface [67, 86]. The method of entrapment isn't immobilized in a direct way on electrodes by addition of a polymer, such as PPy, Chi, agarose, PAni, among others. The method of enzyme entrapment changes according to the polymer, such as the PPy and PAni, since these polymers are deposited on the electrode by electropolymerization. The entrapment of enzymes occurs through the mixture of enzymes and monomers. As opposed to the covalent bonding method, this method shows high stability for enzymes and reduces the leaching besides of to allow the simultaneous immobilized several enzymes, mediators, and additives within the same polymer [67, 87]. Finally, the enzyme immobilization method uses the glutaraldehyde as bifunctional chemical cross-linker for cross-linking protein molecules. This method is formed three-dimensional enzyme complex by covalent bonding through intermolecular cross-linkages among the enzymes. The cross-linking also occurs in combination with other enzyme immobilization methods, being considered a method simple that allows a strong interaction between the biomolecules [67, 88].

Hybrids based on conducting nanostructures and based on carbon and polymers have been used to modify the electrode's geometric area, improving the device's sensitivity. Biosensor is owing to make better the immobilization of the DNA probe and to improve the signal amplification [77]. Some works in the literature describe the development of DNA electrochemical biosensors based on hybrids of conjugated polymers and carbon-based materials. For example, Liu et al. [78] developed a device for the detection of *Mycobacterium tuberculosis* based on IS6110 gene with an electrochemical DNA biosensor. Using a signal amplification strategy and using rGO–AuNPs like a sensing platform and AuNPs–PAni as a tracer element for signal amplification. The rGO was prepared from the exfoliated GO sheets with hydrazine hydrate as a reducing agent. The AuNPs–PAni nanocomposite was prepared from mixed of the distilled aniline with AuNPs previously synthesized from the $HAuCl_4$ solution, with addition dropwise potassium persulfate solution to induce the polymerization. The rGO solution was added on the GCE by dropped. Then, AuNPs were electrodeposited on the rGO/GCE electrode. After blocking the binding site with a hexanethiol solution, the rGO-AuNPs/GCE was incubated with the target DNA solution, and finally, a probe solution with Au-PAni was deposited (Fig. 8I). The characterization of composite showed in SEM images morphological features of the nanomaterials synthesized as large lamellate and scrolled sheet of the rGO besides of

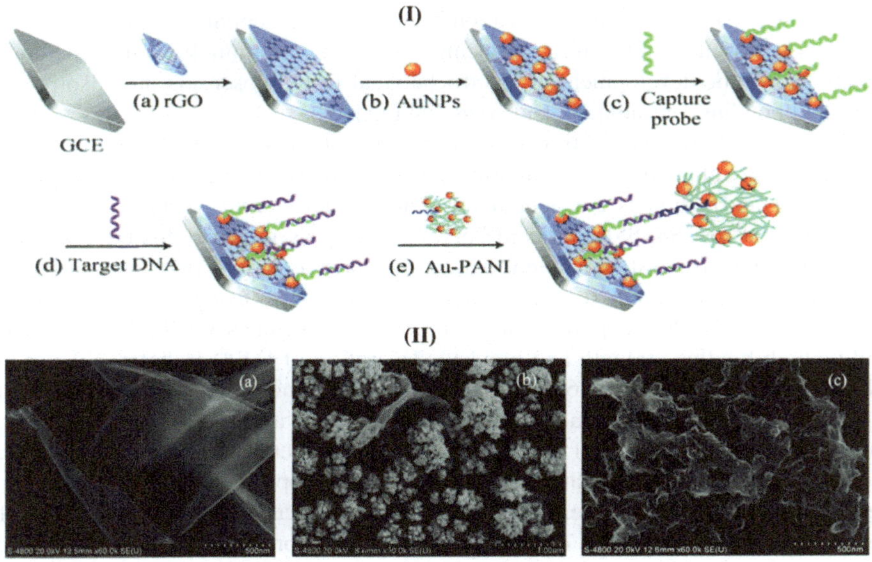

Fig. 8 (I) Schematic of the electrochemical production process: (a) production of rGO membrane; (b) development of AuNP film; (c) immobilization step of captured DNA; (d) combination, and (e) determination of target DNA. (II) SEM images (a) rGO, (b) AuNPs, and (c) Au–PAni [78]

the observed aureatus nanoparticles indicated the AuNPs upon the surface of rGO. After the polymerization of the PAni was observed from the SEM image filiform tubules and AuNPs with PAni indicating the formation of the Au–PAni composite (Fig. 8IIa–c).

The CV and the DPV technique were used, respectively, to characterize and evaluate the analytical performance of the DNA biosensor. The CV different modified in $K_3Fe(CN)_6/K_4Fe(CN)_6$ containing KCl exhibited the oxidation–reduction peak current of $Fe(CN)_6{}^{4-/3-}$ increased after the modification with GO, indicating enhance of electroactive area and device conductivity. AuNPs amplify the peak current, as NPs favor the transmission of electrons. However, by including DNA, the peak current was reduced in cyclic voltammograms due to the blocking of electron transfer by DNA. Furthermore, the results obtained by DPV showed that the peak current grows proportionally with the increase in analyte concentration. The biosensor had a linear range of 1.0×10^{-15} and 1.0×10^{-9} M with good sensitivity, and it can be used in clinical analyses for *Mycobacterium tuberculosis* diagnostics besides can be applied to other pathogenic bacteria.

Shahrokhian et al. [89] developed an electrochemical DNA biosensor based on carboxylic acid PPy (PP3CA) and rGO composite deposited on GCE for the determination of breast cancer gene (BRCA1-associated gene 1). The rGO/GCE was fabricated using the CV technique, immersing the GCE in a GO solution containing lithium perchlorate ($LiClO_4$). After that, rGO/GCE was undergoing the electropolymerization in a solution P3CA and $LiClO_4$ also using CV technique obtaining the modified electrode PP3CA/rGO/GCE. Posteriorly, the linker solution N–Hydroxysuccinimide (NHS) and 1–Ethyl–3–(3–dimethylaminopropyl)–carbodiimide (EDC) in PBS pH $_=$ 7.4 was added on the electrode aiming the activation of the carboxyl groups of the polymerized surface. Finally, a solution of DNA was added to the electrode surface by drop-casting, obtaining the DNA/PP3CA/rGO/GCE. The DNA/PP3CA/rGO/GCE was treated with Ethanoldiamine (EA) in PBS to block residual activated carboxyl groups, and then a solution of the target DNA was deposited on the surface electrode attached DNA to accomplish the hybridization of the surface attached DNA. The morphology of the composite PP3CA/rGO was characterized by atomic force microscopy (AFM) and SEM. The images showed the lamellar structure of rGO nanosheets and fiber-like structures were observed after the electropolymerization process of the PP3CA. The PP3CA/rGO/GCE image was observed, and the rGO sheets acted as a platform in the formation organized of the polymer. The CV, EIS, and DPV techniques were utilized for monitoring the electrochemical behavior of the PP3CA/rGO/GCE. The EIS results using $K_3Fe(CN)_6/K_4Fe(CN)_6$ showed that the diameter of the Nyquist diagram decreased after the activation of the polymer film by NHS/EDC for amine coupling, indicating a reduction of surface electrode negative charge. EIS results also showed an increase in the diameter of the Nyquist after the DNA aminated binds to the carboxyl terminate of PP3CA, indicating that a barrier was formed by DNA agglomeration affecting electron transfer at the electrode/solution interface. After the treatment with ES, the diameter of the Nyquist decreased again due to the EA assisting in removal non-specific attachment of DNA besides of favoring the aligned all chains perpendicular to the surface helping electron transmission. CV results showed PP3CA/GCE that the current responses declined due to immobilized macromolecules to block the diffusion of ions and to minimize current charge in voltammetric responses. Electrochemical results showed that the PP3CA/rGO/GCE presented the signal amplification because occurred the improvement of the electron transfer process. The electrochemical DNA biosensor (PP3CA/rGO/GCE) allowed quantitative analysis of BRCA1 with a LoD of 3 fM, in the range of 10 fM to 0.1 μM using DPV technique. The authors concluded that of electrochemical DNA biosensor fabricated using a GCE modified with PP3CA/rGO composite showed selectivity, sensitivity, repeatability, and reproducibility due to the results showed that the polymer-coated rGO (composite) provides more active sites for the DNA probe immobilization. The DNA biosensor determined trace amounts of DNA target in blood plasma samples obtained the recoveries from 86.3 to 109.8% with relative standard deviations of 3.28–5.47%. The authors concluded that this DNA biosensor can be a new platform for DNA determination in complex matrices.

3 Final Considerations

Hybrid materials applied to biosensors have represented much of the modern advances in analyte detection and bioelectrochemistry. Throughout this chapter, it was observed that the development of CPs and carbon-based nanotechnological materials made it possible to expand the possibilities for manufacturing hybrid materials.

In hybrid materials, the proportions between their components are optimized to mitigate the influence of interfering agents, improve specific properties and generate synergy in biosensors. In association with the sensor element, the works presented here demonstrated that the use of hybrid materials brought gains in selectivity through a decrease in the LoD, improvements in electrical conductivity and signal strength. Electrochemical biosensors also exhibited several competitive advantages as compared to other analytical techniques such as the reduction in the number of reagents or samples required to carry out the analysis, agility in carrying out electrochemical analyses, the possibility of system portability, reduction of waste emission, and cost reduction.

Furthermore, the applications discussed in the fields: environment, food, disease, and DNA show the versatility of biosensors. However, they all have a common factor: They allow monitoring substances that affect the quality of life. In the environment, the detection of emerging pollutants, endocrine disruptors, phenolic compounds, OPs, CBs, etc., favors the monitoring, inspection, creation, and review of norms and laws for the maintenance of well-being and life. In food, this monitoring also contributes to the identification and adequacy of concentrations of compounds that are harmful to the consumer's body or the quality of the food. In diseases, it allows monitoring of clinical pathologies, performing noninvasive or minimally invasive analyses, identifying diseases before the onset of symptoms or asymptomatic, favoring early diagnoses. In DNA, the advances in biosensors allow the rapid and sensitive diagnosis of diseases such as cancer through its biomarkers, because the foundation of accurate and substantiated diagnoses facilitates the treatment of pathologies, contributing to the performance of tests that can be performed at low cost.

Acknowledgments The authors thanks FAPESP (2021/08041-4 and 2017/24742-7) CAPES and CNPQ.

References

1. Canobre SC, Almeida DAL, Fonseca CP, Neves S (2009) Synthesis and characterization of hybrid composites based on carbon nanotubes. Electrochim Acta 54:6383–6388
2. Hur J, Park S-H, Bae J (2015) Elaborate chemical sensors based on graphene/conducting polymer hybrids. Curr Org Chem 19:1117–1133
3. Bourdo SE, Viswanathan T (2005) Graphite/polyaniline (GP) composites: synthesis and characterization. Carbon 43:2983–2988

4. Wang L, Wang D, Zhu G et al (2011) Thermoelectric properties of conducting polyaniline/graphite composites. Mater Lett 65:1086–1088
5. Simoes FR, Capparelli Mattoso LH, Pedro Vaz CM (2006) Conducting polymers as sensor materials for the electrochemical detection of pesticides. Sens Lett 4:319–324
6. Huang J-C (2002) Carbon black filled conducting polymers and polymer blends. Adv Polym Technol 21:299–313
7. Simoes FR, Bulhoes LOS, Pereira EC (2009) Synthesis and characterization of conducting composites of polyaniline and carbon black with high thermal stability. Polim-Cienc E Tecnol 19:54–57
8. Lee H-Y, Jung Y, Kim S (2016) Conducting polymer coated graphene oxide electrode for rechargeable lithium-sulfur batteries. J Nanosci Nanotechnol 16:2692–2695
9. Zhou H, Han G (2016) One-step fabrication of heterogeneous conducting polymers-coated graphene oxide/carbon nanotubes composite films for high-performance supercapacitors. Electrochim Acta 192:448–455
10. Lei W, Si W, Xu Y et al (2014) Conducting polymer composites with graphene for use in chemical sensors and biosensors. Microchim Acta 181:707–722
11. Gokoglan TC, Kesik M, Soylemez S et al (2017) Paper based glucose biosensor using graphene modified with a conducting polymer and gold nanoparticles. J Electrochem Soc 164:G59–G64
12. An HF, Wang XY, Li N et al (2009) Carbon and conducting polymer composites for supercapacitors. Prog Chem 21:1832–1838
13. Albayati SAR, Kashanian S, Nazari M, Rezaei S (2019) Novel fabrication of a laccase biosensor to detect phenolic compounds using a carboxylated multiwalled carbon nanotube on the electropolymerized support. Bull Mater Sci 42:187
14. Yang Y, Zhang L, Li S et al (2013) Electrochemical performance of conducting polymer and its nanocomposites prepared by chemical vapor phase polymerization method. J Mater Sci-Mater Electron 24:2245–2253
15. Branzoi V, Branzoi F, Pilan L, Donisan N (2010) The characterization of some nanocomposites based on conducting polymers and carbon nanotubes obtained by co-polymerization. Rev Roum Chim 55:369
16. Frackowiak E, Khomenko V, Jurewicz K et al (2006) Supercapacitors based on conducting polymers/nanotubes composites. J Power Sources 153:413–418
17. Mottaghitalab V (2005) The influence of carbon nanotubes on mechanical and electrical properties of polyaniline fibers. Synth Met 152:77–80
18. Alqarni SA, Hussein MA, Ganash AA, Khan A (2020) Composite material-based conducting polymers for electrochemical sensor applications: a mini review. Bionanoscience 10:351–364
19. Jain R, Jadon N, Pawaiya A (2017) Polypyrrole based next generation electrochemical sensors and biosensors: a review. Trac-Trends Anal Chem 97:363–373
20. Mohajeri S, Dolati A, Yazdanbakhsh K (2019) Synthesis and characterization of a novel non-enzymatic glucose biosensor based on polyaniline/zinc oxide/multi-walled carbon nanotube ternary nanocomposite. J Electrochem Sci Eng 9:207–222
21. de Araújo GM, Simões FR (2017) Self-assembled films based on polypyrrole and carbon nanotubes composites for the determination of Diuron pesticide. J Solid State Electrochem
22. He BL, Dong B, Wang W, Li HL (2009) Performance of polyaniline/multi-walled carbon nanotubes composites as cathode for rechargeable lithium batteries. Mater Chem Phys 114:371–375
23. Wang Z, Han J-J, Zhang N et al (2019) Synthesis of polyaniline/graphene composite and its application in zinc-rechargeable batteries. J Solid State Electrochem 23:3373–3382
24. Nemade K, Dudhe P, Tekade P (2018) Enhancement of photovoltaic performance of polyaniline/graphene composite-based dye-sensitized solar cells by adding TiO2 nanoparticles. Solid State Sci 83:99–106
25. Cogal S, Ali AK, Erten-Ela S et al (2018) Plasma-based preparation of polyaniline/graphene and polypyrrole/graphene composites for dye-sensitized solar cells as counter electrodes. J Macromol Sci Part-Pure Appl Chem 55:317–323

26. Shin K-Y, Cho S, Jang J (2013) Graphene/polyaniline/poly(4-styrenesulfonate) hybrid film with uniform surface resistance and its flexible dipole tag antenna application. Small 9:3792–3798
27. Thangamani JG, Deshmukh K, Sadasivuni KK et al (2017) White graphene reinforced polypyrrole and poly(vinyl alcohol) blend nanocomposites as chemiresistive sensors for room temperature detection of liquid petroleum gases. Microchim Acta 184:3977–3987
28. Qi Z, Ye J, Chen W et al (2018) 3D-printed, superelastic polypyrrole-graphene electrodes with ultrahigh areal capacitance for electrochemical energy storage. Adv Mater Technol 3:1800053
29. Spinks GM, Xi BB, Truong VT, Wallace GG (2005) Actuation behaviour of layered composites of polyaniline, carbon nanotubes and polypyrrole. Synth Met 151:85–91
30. Qiu YJ, Yu J, Fang G et al (2009) Synthesis of carbon/carbon core/shell nanotubes with a high specific surface area. J Phys Chem C 113:61–68
31. Popov A, Aukstakojyte R, Gaidukevic J et al (2021) Reduced graphene oxide and polyaniline nanofibers nanocomposite for the development of an amperometric glucose biosensor. Sensors 21:948
32. Facure MHM, Mercante LA, Mattoso LHC, Correa DS (2017) Detection of trace levels of organophosphate pesticides using an electronic tongue based on graphene hybrid nanocomposites. Talanta 167:59–66
33. Cesarino I, Moraes FC, Machado SAS (2011) A biosensor based on polyaniline-carbon nanotube core-shell for electrochemical detection of pesticides. Electroanalysis 23:2586–2593
34. Fernandez J, Bonastre J, Molina J, Cases F (2018) Electrochemical study on an activated carbon cloth modified by cyclic voltammetry with polypyrrole/anthraquinone sulfonate and reduced graphene oxide as electrode for energy storage. Eur Polym J 103:179–186
35. Radhapyari K, Kotoky P, Das MR, Khan R (2013) Graphene-polyaniline nanocomposite based biosensor for detection of antimalarial drug artesunate in pharmaceutical formulation and biological fluids. Talanta 111:47–53
36. Zhuang X, Tian C, Luan F et al (2016) One-step electrochemical fabrication of a nickel oxide nanoparticle/polyaniline nanowire/graphene oxide hybrid on a glassy carbon electrode for use as a non-enzymatic glucose biosensor. Rsc Adv 6:92541–92546
37. Rahman MM, Shiddiky MJA, Rahman MA, Shim Y-B (2009) A lactate biosensor based on lactate dehydrogenase/nictotinamide adenine dinucleotide (oxidized form) immobilized on a conducting polymer/multiwall carbon nanotube composite film. Anal Biochem 384:159–165
38. Yang M, Ren X, Yang T et al (2021) Polypyrrole/sulfonated multi-walled carbon nanotubes conductive hydrogel for electrochemical sensing of living cells. Chem Eng J 418:129483
39. Sethuraman V, Muthuraja P, Raj JA, Manisankar P (2016) A highly sensitive electrochemical biosensor for catechol using conducting polymer reduced graphene oxide-metal oxide enzyme modified electrode. Biosens Bioelectron 84:112–119
40. Zheng H, Yan Z, Wang M et al (2019) Biosensor based on polyaniline-polyacrylonitrile-graphene hybrid assemblies for the determination of phenolic compounds in water samples. J Hazard Mater 378:120714
41. Yang Y, Kang M, Fang S et al (2015) Electrochemical biosensor based on three-dimensional reduced graphene oxide and polyaniline nanocomposite for selective detection of mercury ions. Sens Actuators B-Chem 214:63–69
42. Soylemez S (2019) A conjugated polymer and SWCNTs transducer for an effective biosensing tool. J Electrochem Soc 166:B853–B858
43. Nia PM, Meng WP, Lorestani F et al (2015) Electrodeposition of copper oxide/polypyrrole/reduced graphene oxide as a nonenzymatic glucose biosensor. Sens Actuators B-Chem 209:100–108
44. Xu Q, Gu S-X, Jin L et al (2014) Graphene/polyaniline/gold nanoparticles nanocomposite for the direct electron transfer of glucose oxidase and glucose biosensing. Sens Actuators B-Chem 190:562–569
45. Shrestha BK, Ahmad R, Mousa HM et al (2016) High-performance glucose biosensor based on chitosan-glucose oxidase immobilized polypyrrole/Nafion/functionalized multi-walled carbon nanotubes bio-nanohybrid film. J Colloid Interface Sci 482:39–47

46. Radhakrishnan S, Sumathi C, Umar A et al (2013) Polypyrrole-poly(3,4-ethylenedioxythiophene)-Ag (PPy-PEDOT-Ag) nanocomposite films for label-free electrochemical DNA sensing. Biosens Bioelectron 47:133–140
47. Wang L, Hua E, Liang M et al (2014) Graphene sheets, polyaniline and AuNPs based DNA sensor for electrochemical determination of BCR/ABL fusion gene with functional hairpin probe. Biosens Bioelectron 51:201–207
48. Li Y, Zhang Y, Han G et al (2016) An acetylcholinesterase biosensor based on graphene/polyaniline composite film for detection of pesticides. Chin J Chem 34:82–88
49. Liu GD, Lin YH (2006) Biosensor based on self-assembling acetylcholinesterase on carbon nanotubes for flow injection/amperometric detection of organophosphate pesticides and nerve agents. Anal Chem 78:835–843
50. Liu K, Dong H, Deng Y (2016) Recent advances on rapid detection of pesticides based on enzyme biosensor of nanomaterials. J Nanosci Nanotechnol 16:6648–6656
51. Yang Y, Asiri AM, Du D, Lin Y (2014) Acetylcholinesterase biosensor based on a gold nanoparticle-polypyrrole-reduced graphene oxide nanocomposite modified electrode for the amperometric detection of organophosphorus pesticides. Analyst 139:3055–3060
52. Min K, Freeman C, Kang H, Choi S-U (2015) The regulation by phenolic compounds of soil organic matter dynamics under a changing environment. Biomed Res Int 2015:825098
53. Saidur MR, Aziz ARA, Basirun WJ (2017) Recent advances in DNA-based electrochemical biosensors for heavy metal ion detection: a review. Biosens Bioelectron 90:125–139
54. Cui L, Wu J, Ju H (2015) Electrochemical sensing of heavy metal ions with inorganic, organic and bio-materials. Biosens Bioelectron 63:276–286
55. Kaur N, Thakur H, Prabhakar N (2016) Conducting polymer and multi-walled carbon nanotubes nanocomposites based amperometric biosensor for detection of organophosphate. J Electroanal Chem 775:121–128
56. Pundir CS, Chauhan N (2012) Acetylcholinesterase inhibition-based biosensors for pesticide determination: a review. Anal Biochem 429:19–31
57. Kesik M, Kanik FE, Turan J et al (2014) An acetylcholinesterase biosensor based on a conducting polymer using multiwalled carbon nanotubes for amperometric detection of organophosphorous pesticides. Sens Actuators B-Chem 205:39–49
58. Cesarino I, Moraes FC, Lanza MRV, Machado SAS (2012) Electrochemical detection of carbamate pesticides in fruit and vegetables with a biosensor based on acetylcholinesterase immobilised on a composite of polyaniline-carbon nanotubes. Food Chem 135:873–879
59. Viswanathan S, Radecka H, Radecki J (2009) Electrochemical biosensors for food analysis. Monatshefte Chem 140:891–899
60. Kaur N, Thakur H, Prabhakar N (2019) Multi walled carbon nanotubes embedded conducting polymer based electrochemical aptasensor for estimation of malathion. Microchem J 147:393–402
61. Lata S, Batra B, Singala N, Pundir CS (2013) Construction of amperometric l-amino acid biosensor based on l-amino acid oxidase immobilized onto ZnONPs/c-MWCNT/PANI/AuE. Sens Actuators B Chem 1088:1080–1088
62. Lu L, Zhang O, Xu J et al (2013) A facile one-step redox route for the synthesis of graphene/poly (3,4-ethylenedioxythiophene) nanocomposite and their applications in biosensing. Sens Actuators B-Chem 181:567–574
63. Chen C, Xie Q, Yang D et al (2013) Recent advances in electrochemical glucose biosensors: a review. Rsc Adv 3:4473–4491
64. Hsu C-W, Su F-C, Peng P-Y et al (2016) Highly sensitive non-enzymatic electrochemical glucose biosensor using a photolithography fabricated micro/nano hybrid structured electrode. Sens Actuators B-Chem 230:559–565
65. Nambiar S, Yeow JTW (2011) Conductive polymer-based sensors for biomedical applications. Biosens Bioelectron 26:1825–1832
66. He W, Huang Y, Wu J (2020) Enzyme-free glucose biosensors based on MoS2 nanocomposites. Nanoscale Res Lett 15:60

67. Sassolas A, Blum LJ, Leca-Bouvier BD (2012) Immobilization strategies to develop enzymatic biosensors. Biotechnol Adv 30:489–511
68. Xue K, Zhou S, Shi H et al (2014) A novel amperometric glucose biosensor based on ternary gold nanoparticles/polypyrrole/reduced graphene oxide nanocomposite. Sens Actuators B-Chem 203:412–416
69. Tan XC, LI MJ, Cai PX et al (2005) An amperometric cholesterol biosensor based on multi-walled carbon nanotubes and organically modified sol-gel/chitosan hybrid composite film. Anal Biochem 337:111–120
70. Amiri M, Arshi S (2020) An overview on electrochemical determination of cholesterol. Electroanalysis 32:1391–1407
71. Arya SK, Datta M, Malhotra BD (2008) Recent advances in cholesterol biosensor. Biosens Bioelectron 23:1083–1100
72. Dey RS, Raj CR (2010) Development of an amperometric cholesterol biosensor based on graphene-Pt nanoparticle hybrid material. J Phys Chem C 114:21427–21433
73. Ruecha N, Rangkupan R, Rodthongkum N, Chailapakul O (2014) Novel paper-based cholesterol biosensor using graphene/polyvinylpyrrolidone/polyaniline nanocomposite. Biosens Bioelectron 52:13–19
74. Alexander S, Barapeedharan P, Balasubrahmanyan S, Ramaprabhu S (2017) Modified graphene based molecular imprinted polymer for electrochemical non-enzymatic cholesterol biosensor. Eur Polym J 86:106–116
75. Gautam V, Singh KP, Yadav VL (2018) Polyaniline/MWCNTs/starch modified carbon paste electrode for non-enzymatic detection of cholesterol: application to real sample (cow milk). Anal Bioanal Chem 410:2173–2181
76. Chen Y, Guo S, Zhao M et al (2018) Amperometric DNA biosensor for Mycobacterium tuberculosis detection using flower-like carbon nanotubes-polyaniline nanohybrid and enzyme-assisted signal amplification strategy. Biosens Bioelectron 119:215–220
77. Chen Y, Li Y, Yang Y et al (2017) A polyaniline-reduced graphene oxide nanocomposite as a redox nanoprobe in a voltammetric DNA biosensor for Mycobacterium tuberculosis. Microchim Acta 184:1801–1808
78. Liu C, Jiang D, Xiang G et al (2014) An electrochemical DNA biosensor for the detection of Mycobacterium tuberculosis, based on signal amplification of graphene and a gold nanoparticle-polyaniline nanocomposite. Analyst 139:5460–5465
79. Pasinszki T, Krebsz M, Tung TT, Losic D (2017) Carbon nanomaterial based biosensors for non-invasive detection of cancer and disease biomarkers for clinical diagnosis. Sensors 17:1919
80. Tezerjani MD, Benvidi A, Rezaeinasab M et al (2016) An impedimeric biosensor based on a composite of graphene nanosheets and polyaniline as a suitable platform for prostate cancer sensing. Anal Methods 8:7507–7515
81. Asadian E, Shahrokhian S, Zad AI, Jokar E (2014) In-situ electro-polymerization of graphene nanoribbon/polyaniline composite film: Application to sensitive electrochemical detection of dobutamine. Sens Actuators B-Chem 196:582–588
82. Khodadadi A, Faghih-Mirzaei E, Karimi-Maleh H et al (2019) A new epirubicin biosensor based on amplifying DNA interactions with polypyrrole and nitrogen-doped reduced graphene: experimental and docking theoretical investigations. Sens Actuators B-Chem 284:568–574
83. Arora P, Sindhu A, Dilbaghi N, Chaudhury A (2011) Biosensors as innovative tools for the detection of food borne pathogens. Biosens Bioelectron 28:1–12
84. Jesionowski T, Zdarta J, Krajewska B (2014) Enzyme immobilization by adsorption: a review. Adsorpt-J Int Adsorpt Soc 20:801–821
85. Datta S, Christena LR, Rajaram YRS (2013) Enzyme immobilization: an overview on techniques and support materials. 3 Biotech 3:1–9.
86. Cao LQ, van Langen L, Sheldon RA (2003) Immobilised enzymes: carrier-bound or carrier-free? Curr Opin Biotechnol 14:387–394
87. Klotzbach TL, Watt M, Ansari Y, Minteer SD (2008) Improving the microenvironment for enzyme immobilization at electrodes by hydrophobically modifying chitosan and Nafion (R) polymers. J Membr Sci 311:81–88

88. Barbosa O, Ortiz C, Berenguer-Murcia A et al (2014) Glutaraldehyde in bio-catalysts design: a useful crosslinker and a versatile tool in enzyme immobilization. Rsc Adv 4:1583–1600
89. Shahrokhian S, Salimian R (2018) Ultrasensitive detection of cancer biomarkers using conducting polymer/electrochemically reduced graphene oxide-based biosensor: application toward BRCA1 sensing. Sens Actuators B-Chem 266:160–169

Biosensors in Point-of-Care: Molecular Analysis, Strategies and Perspectives to Health Care

Rafael N. P. Colombo

Abstract Nowadays, regardless of the place and situation, it is unmistakable that we are constantly surrounded by sensors of the most diversified types and characteristics. With the advancements of health science, there is an increasing and intense demand for sensors applied to health care to monitor, as frequently as possible, biomarkers of interest, whether in a hospital environment, in a domestic/bedside setting or even with the comfort and elegance of high-tech wearable devices. In the biosensing field, point-of-care (PoC) analyses are becoming progressively appealing in contrast to traditional routes of laboratory assays, and the advancement of nanotechnology with syntheses, characterisations and methods of micro/nanofabrication strongly boosted research in this area during the past two decades. This chapter aims to outline the progress of (nano)materials and biomolecules in PoC sensors, as well as offering readers a concise summary of some of the numerous production methods and processes to devices manufacture, where a few selected works will be discussed, among a lengthy catalogue of excellent and successful ones, in terms of their devices' operational/building strategies. Undoubtedly, the historical evolution of the biosensing field is of utter importance, depicting a road of apprenticeship, and we offer readers a discussed compilation of some temporal milestones from the last decades, from the emergence of the first electrochemical biosensors to the contemporary pressure to incorporating (bio)sensing platforms to smartphones and wearable devices, ubiquitous in modern life. In the light of frontier science, we assess perspectives to the future of PoC biosensing for this decade, where some directions are conspicuous, notably by the foreseeable pathway of technological advances, with accessible and more powerful apparatuses and high-tech gadgets in an environment with the Internet of things (IoT) promoting integration, miniaturisation, dense electronic embedding, automation, real-time data acquisition, digital interconnection and cloud-linked information.

Keywords Point-of-care · Biosensors · Electrochemical sensors · Nanomaterials · Wearable devices · Healthcare

---

R. N. P. Colombo (✉)
Instituto de Química de São Carlos, Universidade de São Paulo, São Paulo, Brazil
e-mail: rafael.colombo@usp.br

F. N. Crespilho (ed.), *Advances in Bioelectrochemistry Volume 3*,
https://doi.org/10.1007/978-3-030-97921-8_7

1 Introduction

Despite the term "biosensor" being of ample usage, for a matter of precision, we can explore its definition as a device that adopts a biologic component (enzymes, antibodies, DNA or RNA, aptamers, cells, among others) to the detection of a chemical analyte [1]. A myriad of sensing methods are existent as optical, thermal and colorimetric, and this chapter will focus on electrochemical biosensors, due to the crescent facility of fabrication, the inherent versatility, high efficiency and low achievable limits of detection.

In this context, many settings and situations have generated a demand to develop analytical quantification techniques performable at the sample collection site, the so-called point-of-care sensors. Many factors contributed to the success of PoC biosensors and the escalating amount of papers on the topic being published every year by the scientific community [2]. Not only do these platforms find their space in personal health care, where PoC devices are most known for, but also in food quality assessment [3, 4], specialised biomedical tests [5], environmental impact evaluation [6], and even illicit drug detection [7].

In the healthcare area, biosensors became a keystone as all sorts of biologic markers can be easily quantified in a laboratory, contributing to high diagnosis power and precision, public safety and improved life quality and expectancy in a world of constantly increasing demands for clinical analysis [8–10]. This kind of platform brings benefits not only to homecare/bedside, probably the most popular scenario of PoC biosensors due to glucometers but also for applications in emergency medical services (EMU) and intensive care units (ICU), where critically ill and premature infants are uninterruptedly monitored, surgery rooms, clinicians' offices and even military operation/training sites [11].

The transition from traditional lab-based biosensors to PoC platforms happened gradually during the past two decades due to a contribution of the following factors:

1. Evolution of nanoscience, with a vast set of easily synthesised nanoparticles [12], carbon nanomaterials [13, 14], and surface nanostructuring either by guided assembly (e.g. sphere-void segment, breath-figure method, self-assembly, and printing) or spatially controlled material removal (such as lithography, ion/chemical etching, dealloying, and thermal decomposition) [15–17];
2. Advances in the incorporation and stabilisation of biological components anchored to surfaces and nanoparticles, including the orientation that plays a crucial role in bioassays' performance, with particular emphasis on enzymes/antibodies [18–21], DNA/RNA [22, 23], and aptamers [24, 25].
3. Technological advances leading to the trend in miniaturised and versatile devices such as lab-on-a-chip sensors [26, 27], which could be engineered to detect a multitude of analytes (multiplex sensing) or perform an initial health screening on patients at the clinician's office level [28].
4. Popularisation of smartphones with increasing integration capabilities to third-party applications, accessories and cloud data storage, emerging as an alternative

pathway to data analysis when coupled to a (bio)sensor platform, often developed to eliminate the need for a trained operator and with immediate processed and interpreted results shown and registered in the smartphone's user screen [29–31].

5. Viability of embedded sensors to wearable devices [32, 33], as smart watches/bands and wrist-patches [34], rings [35], epithelial/skin patches [36], and contact lenses [37].

All these factors contributed to the availability of many PoC biosensors in drugstores, electronic devices with embedded biosensors already being commercialised in a more conscious society seeking palpable and defined metrics of health condition.

By the time of this chapter production, it is worth pointing out that SARS-CoV-2 pandemics further encouraged the trend of local, economical, self-testing. Not only frequent COVID-19 tests became now part of quotidian life and are marketed in multiple general stores, but the isolation and avoidance of public areas postponed regular check-ups, engendering a plight as many individuals with relevant chronic illnesses lack numerical information of their own health status, one drawback that telehealth cannot unravel. This circumstance might soon become a one-off as PoC technologies embark on ordinary routines.

2 A Concise Historical Description

Point-of-care (bio)sensors are on the way to becoming ubiquitous, and one could argue that the past decade of research was decisive to this outcome. While this chapter will discuss some applications of PoC sensors, clever and trending fabrication strategies, and commonplace materials, it would certainly be unfair to not provide readers with a grasp of the rich and fruitful history of biosensors research with representative works that craved a mark to the development of what led to the state-of-the-art PoC devices in the present decade.

When it comes to immunoassays, Dochez and Avery performed one of the most known works of early antigen sensing. In 1917, they demonstrated the detection of a pneumococcal polysaccharide as a product of infection in rabbits' urine, specifically reacting with homologous anti-pneumococcus serum [38]. In the late 50s, Berson and Yalow published the first works of radioimmunoassay focusing on insulin-binding antibodies and I^{131}-labelled insulin, describing kinetic and thermodynamic differences to free and coupled insulin-antibody pair and the consequences of the administration of insulin from diverse animal sources to diabetic individuals [39].

Moving to the 60s, there was an intensification of the usage of biological components to selectively interact with target molecules, and perhaps, the most famous work of the electrochemical biosensing field consists of Clark and Lyons' article, attracted to the idea of continuous monitoring of pO_2, pCO_2 and pH during open surgery processes. The authors elaborated a sensor consisting of a concentrated glucose oxidase (GOx) layer pressed between an inner semipermeable and an outer dialysis membrane. The system was kept in contact with blood, operating in a three-electrode configuration; in their platform, glucose diffuses through the membrane and is converted to gluconic acid, then diffusing towards a pH-sensitive glass resulting in the drop of pH and detected electrochemically with a glucose-pH curve. A similar trend was observed by checking O_2 consumption using hydrophobic membranes [40].

Similar strategies were extensively employed and optimised in the 70s to develop enzymatic biosensors as Guilbault's urea sensor detected by a thin film of urease immobilised in acrylamide gel placed over a cationic/ammonium-responsive electrode; while simple, the biosensor could work continuously for an impressive 21 days with no major loss of activity [41].

In 1974, Mindt and Racine reported a study with eleven electron acceptors to replace oxygen's role in glucose biosensors, as (i) the intrinsic variation of O_2 tension in the blood affects the measured glucose results, (ii) oxygen's limited permeability in biologic fluids could cause "oxygen deficits", once again reflecting in differences in O_2 tension and (iii) high electrode potentials are required for hydrogen peroxide reaction, cumbersome yet natural characteristics of the matrix; their work opened a pathway to mediated biosensors—the second generation of enzymatic sensors—with hexacyanoferrate (III) as a recommended mediator [42].

Discussions on the third generation of enzymatic biosensors, with direct electron transfer (DET), were also seen in the 70s with Berezin, Scheller and Varfolomeev's works suggesting the viability of a pathway where electrons hopped from the active site of an enzyme closely anchored to the respective electroactive surface, commonly Hg or amalgamated metals [43]. It is worth stressing that DET and immobilised enzymes' structure effects, even 40 years later, still present as a controversial topic [44–46].

It is valuable reminding that still in the 70s, Engvall and Perlmann presented the Enzyme-Linked Immunosorbent Assay (ELISA) using Immunoglobulin G (IgG) from rabbit serum to promote antigen-antibody interaction [47]; ELISA became one of the most used immunoassays to date.

The subsequent decade was marked by additional studies of immunosensors as those of Liedberg, Nylander and Lunström [48], and the incipient shift of interest to smaller or implantable biosensors, arguably marking the embryonic stage of PoC sensing. In 1982, Shichiri and co-workers developed a GOx-based, implantable, needle-type glucose biosensor, less than a millimetre thick and able to monitor in vivo glucose levels for up to seven days [49].

Ferrocene and derivatives were found to fulfil requirements for usage as a mediator to glucose monitoring [50], and in 1987, Matthews and co-workers presented the first pen-sized glucometer for personal glucose control, Fig. 1, marketed by "MediSense Inc" under the name "ExacTech", using ferrocene as one of the mediators and glucose dehydrogenase (GDH) as the active enzyme in a disposable test-strip [51]. The device had tremendous success, causing a revolution in the health care of diabetic patients and, although decades ago, one should notice that modern glucometers do not differ widely from theirs in terms of the general strategy. Readers with additional

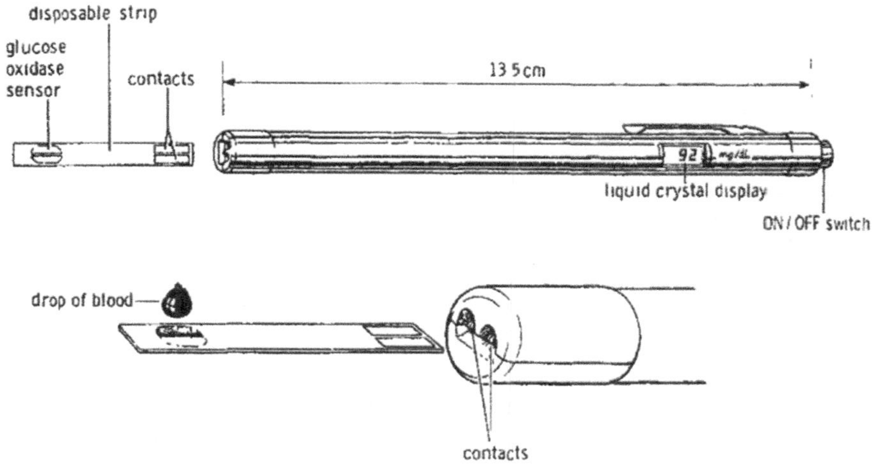

Fig. 1 Representative schematic of the first pen-sized glucometer for personal blood glucose level monitoring. Ferrocene and GDH were used as mediator and active enzyme. Reproduced from [51] with permission of the publisher

interest in glucometers history can find comprehensive studies of the last 40 years of development [52, 53].

The late 80s was also marked by the Campbell, Wagner and O'Connell patent of solid-phase assays [54], an evolution from paper chromatography that found vast applications in the following decades.

It took another decade to the marketing of the first in vivo glucose biosensor by Minimed which, however, did not provide real-time data as three-day glucose level data was registered and stored to be extracted and analysed in a physician's office [53]. Cygnus Inc., in the early 2000s, was granted the first FDA approval for a transdermal glucose monitoring system, operating by reverse iontophoresis and advertised under the famous name of GlucoWatch Biographer.

Research groups in the 90s found themselves in a scenario with a diversity of analytical techniques, surface functionalisation protocols with the large availability of biological components and the certainty of commercial interest of disposable and portable biosensors, which explains the upsurge of a large number of whole-blood point-of-care devices for a multitude of analytes as creatinine, urea, pCO_2 [55], pH, haematocrit and haemoglobin, pO_2, electrolytes, urea and lactate, troponin, creatine kinase and coagulation profiling, often operating in a multiplex mode and providing results in a few minutes [56, 57].

Since it became evident the viability and commercial success for these sensors for both personal bedside healthcare and specialised hospital analysis, the 2000–2010s period was marked by a growing focus on platforms inexpensiveness [58] and disposability [59], smart-materials and smart-devices, the dissemination of the "lab-on-a-chip" idea [59] and multiplexing capabilities [28, 60].

Finally, the past decade surrounded the construction of platforms with greater sensing power and complexity, now viable due to evolution in microfabrication technologies [61], leading to miniaturised portable devices [26], integration and smartphone-based biosensing [30, 31].

3 Applications

3.1 Personal or Bedside Care

The impact of a disease on an individual is weighted by the discomfort and disabilities caused, with repercussions to health service and productivity in the economic sphere. The leading illnesses in the modern world are coronary ischemic disease and cerebral hypoperfusion, with infectious diseases still causing disabilities and high mortality in low-income countries [11].

Point-of-care technologies empower users to monitor their own health while also endowing privateness of personal testing.

In a world in which wearable devices such as smartwatches are becoming commonplace, often with embedded heart rate and pO_2 sensors, academics are producing highly useful electrochemical sensors: multiplexed electrolyte, glucose and lactate in sweat [62], cortisol [63], cholesterol [64] and even bacteria in the tooth enamel [65]; the continuous monitoring of major physiologic markers could be intrinsic for improving life quality as abnormalities could be early noticed, quickly directing the individual to the appropriate medical support.

A recent publication by Yoon and co-workers described an ingeniously constructed multifunctional skin-worn patch, as shown in Fig. 2. The patch operates via four transducer micropads: the electrocardiography (ECG) and temperature are measured by micropatterned gold spiral and platinum array, respectively; pH is quantified by a polyaniline (PANI)-coated gold electrode relative to a polyvinyl butyral (PVB)-coated Ag/AgCl reference electrode, and glucose is detected by GOx onto reduced graphene oxide [66].

PoC devices aiming to aid chronic diseases carriers are also part of recent efforts; researchers developed a test-strip to monitor intestinal alkaline phosphatase, a biomarker useful to assess the severity of inflammatory bowel disease and colitis. The user-friendly assay is compatible with glucose meters and delivers the quantification in 15 min [67]

Being able to confidentially perform home-based analyses can also be beneficial to detect conditions such as sexually transmitted diseases (STD) without causing users possible embarrassment. Xu and co-workers developed a sensitive PoC electrochemical biosensor with capture and label probes capable of promptly detecting HIV DNA fragments; the analysis can be performed by the individual's personal

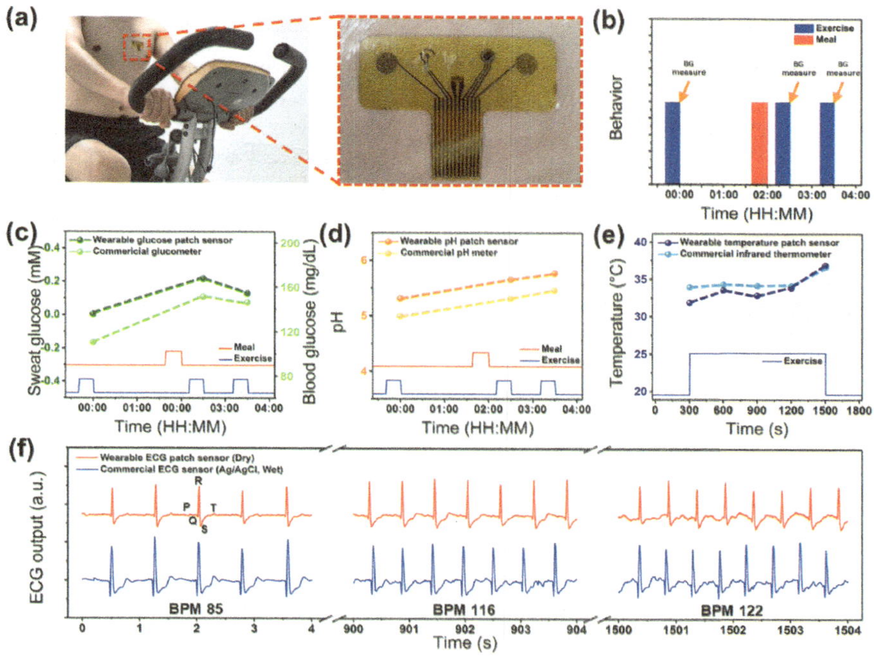

Fig. 2 **a** The multifunctional skin patch attached to the chest area; **b** timetable of glucose (BG) measurements; **c** corresponding measured sweat/blood glucose, **d** pH, **e** temperature and **f** registered ECG profiles. Reproduced from [66] with permission of the publisher

glucometer [68]. Electrochemical genosensors for the Herpes virus can detect intermittent shedding events of the virus in urine samples, important for the high-risk population as immunocompromised patients [69].

3.2 *Physician's Office Testing*

A simple yet efficient paper-based PoC electrochemical sensor was developed by Tseng and co-workers to detect microalbuminuria in urine, an important marker for early detection of kidney disease in patients. The cardboard strip test containing an active modified filter paper coupled to screen-printed electrodes, as shown in Fig. 3, worked in real urine samples, a test that could be easily and quickly performed in the clinician's environment [70].

Ordinary markers of inflammation as C-reactive protein and interleukin-6 can be detected in 5 min via a portable detection unit, collecting material in urine through a dipstick [71], same for creatinine quantification [72]. More specific clinical hallmarks as transforming growth factor-beta 1 (TGF-β1), delved into for renal disease

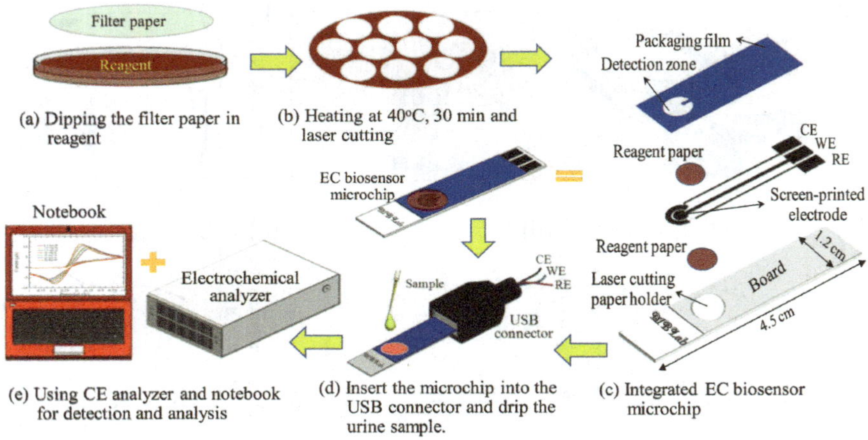

Fig. 3 Schematic of the albumin sensing microchip fabrication, constructed using filter paper, silver/carbon-printed electrode, a dye and a redox mediator. The chip is operated via a portable μ-potentiostat. Reproduced from [70] with permission of the publisher

assessment, can also be quantified in the clinicians' office through urine sampling [73].

Electrochemical PoC tests are also available to pre-evaluate non-STD infection, either by recognising DNA/RNA or hallmarks of infection (e.g. polyuria) [74]; these advances are not only worthwhile by dismissing highly skilled professionals but also by shortening analysis time from days, as the traditional route of culture and polymerase chain reaction, to less than an hour [75].

A complete screening for STDs is another aim, with some PoC electrochemical sensors developed for usage by the patient itself, as for HIV testing, and others requiring a serum extraction and less trivial interpretation as syphilis tests [76].

Other works, as Uludag and co-workers, employed the same idea by developing a biochip with an embedded microfluidic system capable of detecting cancer biomarkers as blood serum flows onto the enzyme-modified substrate, with good results and fast analysis time; the processing device was developed so that a diversity of markers could be detected with distinct biochips [77]. These initial screenings could benefit patients as the clinician can deeply assess one's health condition at the very first appointment with precise and objective biomedical data on hand [11].

3.3 Hospital and Surgery-Room Level Monitoring

When it comes to ICU, EMS or surgery environments, current methods of analysis that require extraction of blood samples and submitting to a laboratory to later receive the results poses a problem as in many circumstances, the concentration of the analyte of interest can change in a matter of minutes, therefore requiring the quantification as fast as possible and ideally in a real-time fashion.

A typical example of a situation requiring rapid identification of the patient's condition to provide immediate medical reaction consists of drug abuse, a common but critical issue as World Health Organisation estimates ca. 275 million illicit drug users worldwide and responsible for 18 million lives lost [78]. Point-of-care devices can carry out fast drug analyses, as Hassan and co-workers showed with a microfabricated disposable potentiometric sensor for benzodiazepines in plasma, urine, saliva and human milk [79].

Recently, Tanak and co-workers addressed the issue of parathyroid hormone (PTH) measurement in hyperfunctioning adenoma surgeries. Due to its ca. 3 min half-life, PTH is a vital tool to detect whether the resection was successful, yet demanding to be quantified rapidly during surgical progress instead of the traditional 60-min route, where surgeons wait for the result, and the patient sometimes needs complementary surgeries in the following weeks. In the aforementioned work, a PoC biosensor based on an antibody-modified electrode strip was used, and by electrochemical impedance spectroscopy (EIS), the quantification can be performed in the surgery room with results in less than 15 min [80].

3.4 Field-Testing: From Food Safety Assessment to Military Hazards' Scrutiny

Field-testing of livestock plays a crucial role to support high productivity, in which the main ambition consists of the early diagnosis of pathogens without extensive veterinary infrastructure or costly equipment. Gattani and co-workers recently reviewed the progress of PoC biosensors for veterinary—especially bovines—care; the rich literature comprises biosensors for pathogens like E. coli, Listeria, Brucella, Campylobacter, Clostridium and diseases such as mastitis and ketosis [81].

When it comes to military and security affairs, biothreat agents are one of the main worries that require fast and sensitive detection procedures to be performed at the site, once again demanding PoC biosensing platforms, ideally with multiplexing capabilities; pressure for biosensors development sharply increased after 2001s anthrax-laced letters in the USA—the Amerithrax case [28]. In this context, electrochemical sensors chips have been developed to sense pathogenic microbial cells and spores in vitro and in situ, either in biologic fluids or supplies like food and water, providing users with a quick and reliable screening rationale to the safety of consumption or already existing health conditions [82, 83]. Biodefense samples can be analysed *on-site* in 16 *min* for the presence of influenza A, *Bacillus anthracis, Yersinia pestis* and *Francisella tularensis* [84].

3.5 Neglected Diseases Screening

PoC biosensors not only achieved their spotlight because of local analyses capability and the speed in displaying the results, but also for offering light in scenarios such as neglected diseases in countries with reduced medical infrastructure. In these regions, the lack of equipment and supplies for biologic material collection and analyses generates a panorama of illness, suffering and severe economic and social consequences, as described in a comprehensive review by Deroco and co-workers [85].

In these places, the possibility to perform fast analytical runs with portable, low-cost devices employing disposable test-strips drastically raises the chances of therapeutic success and better prognosis, ameliorating the life quality of the population.

Since EIS is extremely sensitive to electrode-solution interface changes and often label-free, several neglected diseases sensors rely on impedimetric measurements, as Garrote's biosensor for Dengue through NS1 protein biomarker detection, which can provide results in a matter of minutes in a pre-fabricated electrode [86]. Similar strategies resulted in the recent development of PoC sensors for Malaria [87, 88], schistosomiasis [89], Chikungunya [90], leishmaniosis [91] and Zika virus [92]. Field-effect transistors (FET), although still seldom applied for neglected diseases, will probably find a larger space in literature soon, as successful applications to detection have been reported, e.g. Ebola [93, 94].

3.6 Recent Miscellaneous of Tests: From COVID-19 Sensing to Forensics

SARS-CoV-2 pandemics generated a restraint to the free movement of people, making mandatory frequent testings for indoor activities, even after a broad vaccination coverage.

Notwithstanding the challenges for accurate diagnosis of SARS-CoV-2 and interpretation of tests' results [95, 96], point-of-care tests were promptly reported as a cheaper and faster alternative to RT-PCR, aiding the early identification of infected individuals, hence preventing the virus' spread.

Torres and co-workers developed a low-cost, yet powerful swab-based test operating via impedimetric mode. Authors collected nasopharyngeal/oropharyngeal or saliva samples and applied these to screen-printed electrodes functionalised with angiotensin-converting enzyme-2 (ACE2)—the recognition agent—and covered with BSA for anti-biofouling, as represented in Fig. 4, achieving good accuracies, specificities and sensitivities [97].

Methods based on portable electrochemical analysers were also developed for forensic/criminal affairs, with examples of chemically modified paper-based chips able to detect ketamine, a date-rape drug, in alcoholic and non-alcoholic drinks [98];

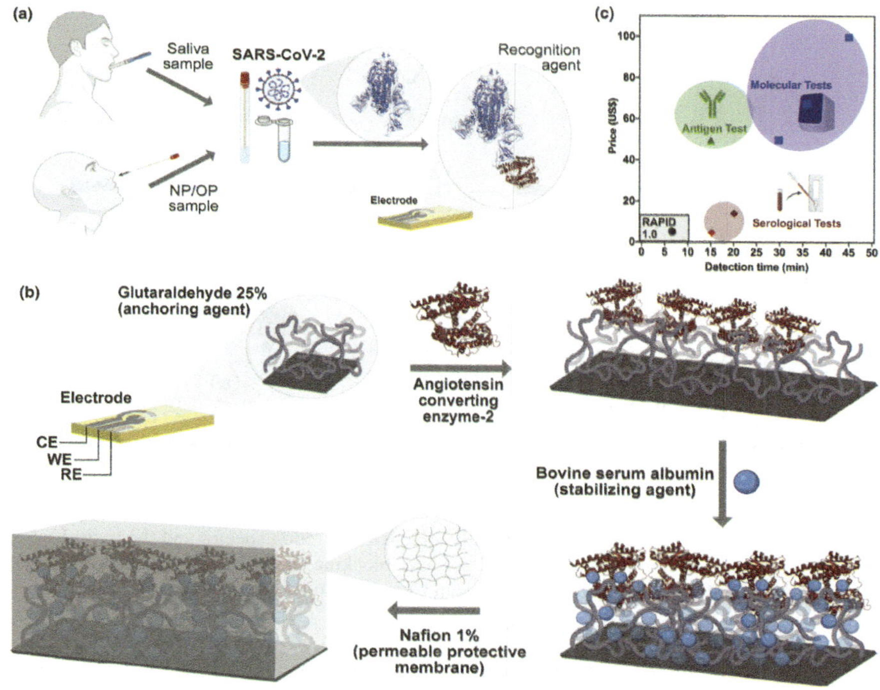

Fig. 4 Low-cost SARS-CoV-2 PoC biosensor operation: saliva or naso/oropharyngeal fluid is collected, and the biosensor electrochemically detects the specific spike-protein to ACE2 binding events through impedance variations. Reproduced from [97] with permission of the publisher

a similar approach led to a portable device for *on-site* scopolamine identification in beverages [99].

4 Strategies, Materials and Surface Engineering

4.1 Electrode Materials

Pioneering biosensors employed expensive materials such as metallic electrodes (Au and Pt mainly), glassy-carbon or ITO-glass electrodes. As new properties became important and PoC now focuses on disposability and integration, cheap materials such as paper gained importance, with the additional benefit of mechanical flexibility and lightness that is particularly desirable for designing wearable patches, and the ubiquity for scaling up. As a fibrous matrix, the paper provides useful inherent characteristics as the wettability that drives homogeneous aqueous phase spread onto the entire surface, the dense disposition of the fibres that allow pre-loading the

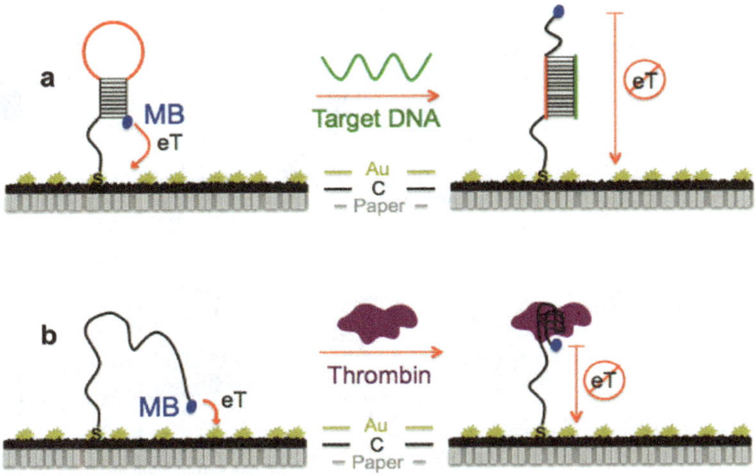

Fig. 5 Representation of a typical ssDNA/aptamer-based sensor operating in an ON-to-OFF configuration; the distal end of the probe is covalently attached to a redox species and the hybridisation causes the reporter to move away from the surface, blocking electron transfer. Reprinted with permission from [102]. Copyright 2014 American Chemical Society

material with chemicals, biocompatibility of the cellulose and friendly disposability [100].

Among the first works concerning electrochemical detection with paper-based PADs, Henry's group showed uric acid, glucose and lactate sensing [101]. In the last decade, several advances were achieved and successful paper-based sensors have been increasingly reported. Among those, the example of Crooks' sensor for DNA and Thrombin [102], based on conformational switching as schematised in Fig. 5.

Of course, the miniaturisation and portability of microfluidic paper-based analytical devices (μPAD) culminated in the viability of homemade simple electrodes with even no modification of the paper, as demonstrated by Kubota and co-workers [103].

Carbon nanomaterials such as graphene and carbon nanotubes (CNT) are also valuable due to their high conductivity and specific surface areas. The drawback of graphene's troublesome functionalisation often requires the usage of its oxidised form, graphene oxide (GO) [13]. Several processes for GO chemical functionalisation are described in the literature [104], and researchers often find the need for a following reduction step to restore part of the sp^2 character of the material, increasing the conductivity.

Gonçales and co-workers studied the effects of GOx immobilisation onto a carbon felt decorated with free-standing reduced graphene oxide (RGO) sheets and cup-stacked CNTs, as shown in Fig. 6. Results indicated that GOx suffers conformational changes when adsorbed on the surface of CNTs, especially RGO, although not releasing FAD moieties; interestingly, improved kinetic parameters were observed in

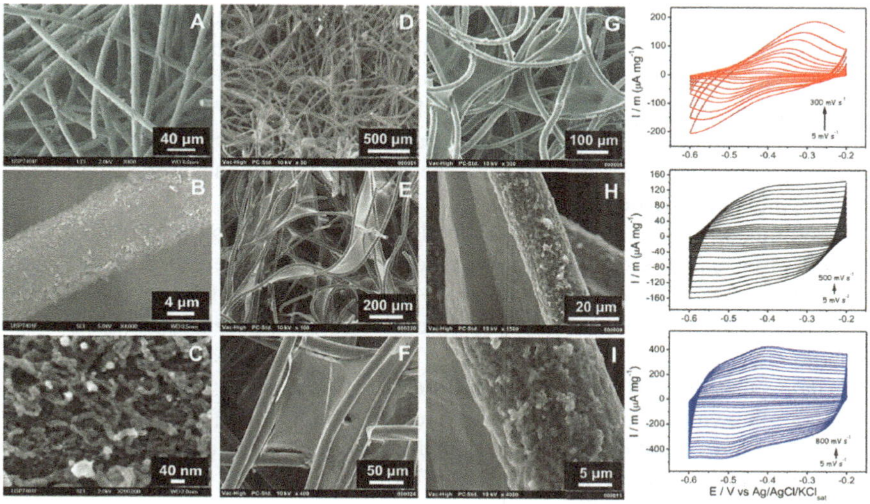

Fig. 6 Carbon felt electrode decorated with (A, B, C) cup-stacked CNTs, (D, E, F) RGO and (G, H, I) CNTs and RGO. After adsorption of GOx, voltametric profiles indicate improved kinetics for CNTs (red CV) and RGO (black CV), but an even greater effect in the composite matrix (blue CV); although electron transfer is improved, GOx is not in the native conformation anymore. Reproduced and modified from [46] with permission of the publisher

the composite matrix compared to RGO only or CNT only, which exposes a particular interaction of the RGO and its contact points in the edges of CNTs around the felt fibres.

Graphene and CNTs received further attention as a PoC electrode material of choice due to providing high conductivity—to screen-printed electrodes, in particular—while conferring chemical inertness, biocompatibility, optical transparency, flexibility and mechanical resistance.

Low and co-workers developed a biochemical sensor for microRNA using a screen-printed carbon electrode (SPCE) and GO, which was electrochemically reduced to graphene in a composite with gold. ss-DNA was chemically anchored to the electrode via thiol chemistry and responded as the recognition probe [105].

The advances in the FET technology also pushed graphene incorporation to the so-called GFETs, and successful examples of portable, label-free GFETs are present in literature, as Wang and co-workers GFET aptasensor for lead quantification in human blood [106].

Nanoparticles are also one of the most common elements of biosensors, simply due to the countless tailorable properties. Among them, the design of size, shape and composition before the materials' syntheses. Nanoparticles in biosensors provide conductivity, high surface area, bridges for specific binding of molecules and even as labels to antibodies.

In one of the approaches, a DNA PoC biosensor was engendered using a folding paper technique, and graphene was added as a conducting layer; AuNP were immobilised to graphene's surface and served to both increase conductivity and as anchorage spots to a thiolated linker. An ssDNA was responsible for the interaction with the target analyte [107].

Among commonplace nanoparticles, gold-coated magnetic nanoparticles can act in a particular way as they can be functionalised with biologic components as antibodies and increase accessibility to the target and later being concentrated to a surface through the magnetic field [108].

The embedded electronics also implied the higher usage of silicon, quartz and glass in chips and boards, which more recently began to be replaced by polymeric materials that have two main advantages, lower cost and surface tuneability, as a diversity of surface properties can be achieved and are therefore interesting to surface-analyte interaction.

Among them, PMMA [109], PET, PS and PC are widely present in literature, with long known protocols for manufacture and modification [110]. More recently, copolymers have brought further attention due to the even greater range of properties and surface interactions to biomolecules [111].

4.2 Sensing and Recognition: Enzymes, Mediators, DNAs, Aptamers and Antibodies

To prevent non-specific interactions and achieve higher sensitivities, with a high S/N ratio and low limit-of-detection, biosensors rely on biologic components, most commonly enzymes, antibodies, DNA/RNA and aptamers.

Enzymes are still the most employed elements in biosensors—"catalytic biosensors"—and efforts have been made since Clark's 1962 inaugural glucose biosensor to (i) widen the applications using a whole set of enzymes and strategies and (ii) deal with the main challenges of this biomolecules class: chemical and thermal stability. Table 1 shows some of the main classes of enzymes and their major roles.

The long and successful journey of the first generation of enzymatic biosensors led to the production of many biosensors in previous decades, since oxygen abundance in fluids acts as a simple mediator, however, implying in its concentration dependence, high potentials for peroxide redox reactions and being prone to chemical interferents. The discovery of chemical mediators as ferrocene deeply improved many of the main drawbacks of the first generation, still, only glucose sensors achieving complete commercial success. Approaches on the third generation of biosensors eliminate the mediated step and are part of the frontier research.

It is evident that the success of enzymatic biosensors—and the PoC devices based on these biologic components—depend on the viability to large scale production and, therefore, the stability in the resultant test-strip for weeks. Stabilisation techniques are being developed especially with encapsulation, covalent-anchoring to the electrode surface, cross-linking and hydrogels layering [112].

Table 1 Selected examples of enzymes' classes, subdivisions, examples and their catalytic roles

Class	Subclass	Example	Typical Function
Oxidases	O_2 acceptor, acting on CH-OH of donor	(1) Glucose oxidase/FAD coenzyme (2) Alcohol oxidase/FAD coenzyme (3) Alcohol dehydrogenase/NAD^+	Catalytic oxidation of the substrate using O_2 (ADH can catalyse the reverse reaction)
	Acceptor dependent, redox acting on CH-CH	(1) 5α-Reductase/NADPH cofactor (2) Bilirubin oxidase/multicopper-centre	Androgen/oestrogen metabolism, interconverting steroids; oxidation of bilirubin, part of porphyrin and chlorophyll metabolism
	NAD/O_2 acceptor, acting on $CH\text{-}NH_2$	(1) Monoamine oxidases/Flavin-based (2) Amine oxidases/Cu^{2+}-based	Oxidative deamination of monoamines
	Peroxide-acting enzymes (Peroxidases)	(1) Catalase/Fe-heme groups (2) Horseradish peroxidase/heme group	Redox conversion of peroxides (including organic)
	Oxygenases	(1) Luciferase/flavin mononucleotide (2) Nitric oxide synthases/NADPH, FAD, FMN, heme (3) Cyclooxygenases /heme-group (4) Cytochrome P450/Fe-heme	Oxidation of substrate by transferring of oxygen from molecular O_2
	Other N-donor	(1) Nitrate reductase/NAD acceptor (2) Nitrite reductase/cytochrome acceptor, possible Cu-containing	Redox reduction of nitrate to nitrite; nitrite to several products
Hydrolases	Phosphatases	Alkaline phosphatase/cysteine, zinc and magnesium dependent	Dephosphorylating biomolecules
	Linear amide hydrolases	Urease/nickel-dependent	Catalytical hydrolysis of urea to carbamate (spontaneously hydrolysed to ammonia and CO_2)
	Carboxylic ester hydrolases	Acetylcholinesterase	Catalytical hydrolysis of choline esters

Another class of commonplace biologic components is antibodies, Y-shaped serum glycoproteins associated with defence mechanisms of living organisms—identification and targeting—and with wide usage in biosensors as recognition elements due to their intrinsically high specificity to their kindred antigen [113]. Immunoassays such as ELISA, immunofluorescence, Western Blot and electrochemical immunoassays (EIA) all employ antibodies and dominate the clinical analyses field, with success to both analytical performance and chemical stability.

Electrochemical PoC biosensors using antibodies—“immunosensors”—focus on three main practical aspects: (i) immobilisation of the antibody to the electrode surface, with appropriate geometrical orientation; (ii) accessibility to antigens, improving limit-of-detection, eventually with new technologies as disposable electrodes [108], and (iii) protection against non-specific adsorption, with antifouling strategies [114]. Several approaches are present in literature, with random adsorption, avidin-biotin specific interaction and covalent linkage of antibodies through amine, thiol or aldehydes [113]. Examples of antibody-based PoC sensors can be found throughout this chapter.

We also briefly present examples of DNA/RNA-based PoC biosensors, where ss-DNA microarrays in the biorecognition interface are responsible for the selective interaction to complementary strands, leading to the identification of genes [115]. The conception of these devices goes through (i) the immobilisation of the designed strand to the electrode, through bare electrostatic interaction that presents drawbacks of complex hybridisation dynamics, or chemical-binding, especially through thiolated ssDNA, justifying the large number of papers also employing Au electrodes or AuNPs. (ii) The strategy for signal transduction, in which one option consists in labelling with electroactive probes either exploring affinity to grooves in the hybridised or ss-form of DNA or employing ss-DNA chemically coupled to electroactive species as ferrocene and exploring differences of freedom before/after hybridisation with the consequence to electrochemistry in both forms [116], as already shown in Fig. 5. The option of label-free most often relies on electrooxidation of guanidine in the complementary strand, although exploring surface accessibility to redox probes before/after hybridisation is another possibility.

The last class of interest consists of aptamers, artificial oligonucleotides/peptides designed to bind specifically to target molecules, ions and even cells. Their usage in biosensors—“aptasensors”—is similar to antibodies, despite the higher versatility to a broader range of analytes and ease of functionalisation/labelling [117]. Thrombin is by far the most common target for aptasensors in literature, playing the analogue role as glucose for enzymatic sensing; labelling with ferrocene, methylene blue or anthraquinone is the dominant approach, with other less-popular options as coupled-enzymes and NPs also available. Strategies for label-free discussed for antibodies-based biosensors are also employed to aptasensors, exploring accessibility to redox-active species in solution to the electrode surface as aptamers interact with the target [118].

4.3 Manufacturing Strategies: Immobilisation, Screen-Printing, LAE, Embossing, Hydrogels, and Antibiofouling

Strategies to PoC biosensors fabrication can be thought to follow four main stages: (i) design of the electrode and choice of the manufacturing method, (ii) selection of the biologic component and the method for surface immobilisation, as well as auxiliary elements for transduction as redox probes, detection/labelled-antibodies, and NPs, (iii) additional steps to increase the stability of the biomolecule and prevent non-specific adsorption, and (iv) engineering and integrating to power sources, portable μ-potentiostat units and wireless technologies, whenever applicable.

Electrodes can be prepared by a multitude of methods, from simple printing processes to complex yet powerful lithographic ones, and authors should keep in mind that costly and hard to upscale processes might lose importance over others requiring few steps and easy to mass-produce. The choice should also consider whether simple flat or patterned electrodes for microfluidic applications are the objectives.

Some of the prevailing methods are shown in Table 2, with their descriptions and practical observations.

Figure 7 represents the wax-printing as a particular case of electrode-printing method, creating a hydrophobic barrier onto carbon or paper-based underlying substrates.

The choosing and immobilisation methods for biologic probes are a vast topic, and readers are referred to the previous section for a brief discussion.

Strategies for bioassay stabilisation include the association of hydrogels; this class of materials can be used as a matrix or interface between the solid electrode's surface and the protein or entrapping the enzymes. This provides a hydrophilic environment with increased loading capacity and tuneable water content for enzymes' long-term stabilisation, thus avoiding the inherent lability and tendency to denaturation in bare contact to solid interfaces [124].

Not less importantly, prevention of non-specific adsorption is extremely valuable to the application of PoC biosensors to real samples with multiple and concentrated components as albumin in blood; traditional methods to prevent loss of activity by biofouling employed a BSA or PEG-SAM layer, which, however, also hinders electron transfer reducing the device's conductivity. A zwitterionic layer has also been proposed as a valid approach to avoid high built-in impedance. Only recently, a simple and robust antifouling coating was reported as being able to significantly prevent non-specific adsorption for up to a month when exposed to unprocessed human plasma [114].

Table 2 Collection of the most employed methods to electrode fabrication for PoC biosensors, along with the description of involved steps and a balance of advantages/disadvantages

Methods	Description	Typical materials	Advantages	Disadvantages	Obs	References
Screen-printing	Electrodes are produced by printing an ink/paste to a plastic, paper, or ceramic base; Mediators, NPs, and biomolecules can be added to the paste	Carbon ink, Pt/Ag paste Graphene, CNTs often included	Low cost Scalable to automated (high production rates) Versatile	Susceptible to batch-to-batch variations Restricted to flat substrates	Very successful approach to disposable chips due to the low cost and automated production	[119]
	Wax-printing: Wax layer is printed through a cutter printer and transferred to underlying substrate by a heat press	Substrates: Carbon/SPCE and metallic	Efficiently patterning of electrodes with a hydrophobic layer Low cost Scalable method	Limited resolution	Handy to limit spreading area of most PoC aqueous matrices; Thermal step may be inviable to electrodes pre-loaded with biomolecules	[120]
Photolithography (and similarly scanning beam lithography)	Photoresist coating and baking forms a film onto the substrate, exposure through a mask patterns the resist; etching steps transfer the pattern to the underlying substrate	Si or SiO_2, metals as Au and Pt and polymers as PET	Precise and reproducible patterning at submicron scale	Multiple steps required Expensive equipment and infrastructure of clean rooms Performance issues on non-flat substrates	Successful applications to multi-channel, flowing-sample chips	[121]

(continued)

Table 2 (continued)

Methods	Description	Typical materials	Advantages	Disadvantages	Obs	References
Spin-coating and Flow-coating	The coating substance is dissolved in a volatile solvent, spread to the entire surface and either the substrate is spun or a gas stream flows through the substrate	Substrate: Si or SiO_2, metallic, polymeric materials (not soluble to the employed solvent) Coatings: polymers, colloids, NPs and graphene oxide	Low cost Rapid production and scalable to large substrates	May suffer large batch-to-batch variations Can suffer from radial heterogeneity	Adaptable to produce single/multi-layered porous films	[122, 123]
Stamping, embossing and moulding	Hard-pattern: (i) Step-and-flesh: Rigid mould covered with photocurable solution and exposed to UV (ii) Nanoimprint: Rigid mould pressed against a thermoplastic polymeric layer and heated Soft Pattern: master patterned mould transferred to PDMS and replicated	Silicon and quartz as rigid moulds PDMS as mould/stamp Polymers in general as stamped layer	Cheaper method for pattern-transferring over large areas	Require multiple steps Time consuming processes Clean rooms recommended	Useful processes for unconventional nanofabrication; Requires careful attention to mould fouling and thickness variations	[61]

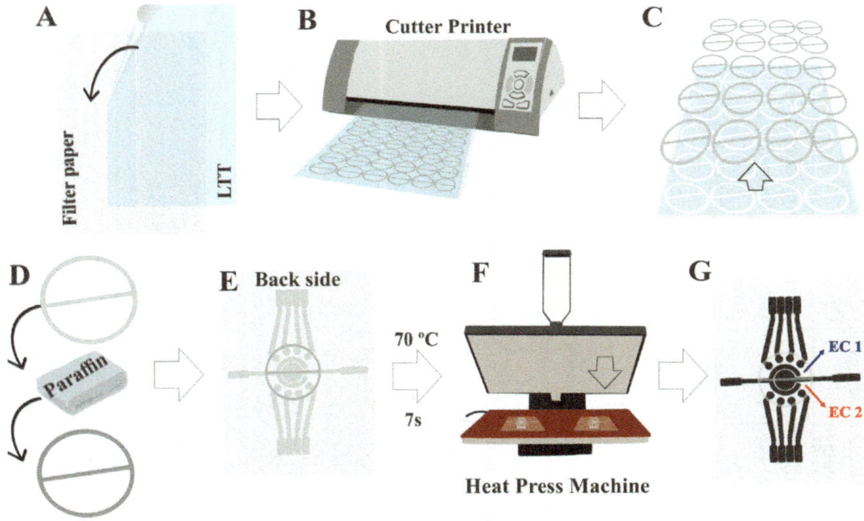

Fig. 7 Schematic of the wax-printing process; filter and transfer paper sheets are coupled (**A**) and a pattern of the electrodes is cut (**B**) accounting for the number of electrodes required (here, 8 WE + RE + CE). The transfer-mask is dipped into molten wax (**D**), aligned onto the paper electrode (**E**) and heat-pressed (**F**), resulting in the desired electrode geometry (**G**). Reproduced from [120] with permission of the publisher

4.4 Technological Advances: Miniaturisation, Integration, Automation, and Multiplexing

The trend to smartphone-integrated sensors becomes patent as the mark of ca. 6.4 billion smartphone users worldwide is hit in 2021, accounting for 80% of the global population; for the first time, half the world's population has access to mobile Internet of increased speed and coverage, as 4G catches up with 3G availability. GSMA reports further stress a significant drop in the cost to entry-level, Internet-enabled smartphones relative to average monthly incomes, and so is the cost for mobile access [125, 126]. Projections point out that by 2025 almost 75% of all Internet users will access it solely through mobile devices, representing a scenario where, even in low-income countries, portable devices with integration capabilities to sensing accessories could provide additional tools to public health care.

When it comes to smartphone-based PoC technologies, Sun and Hall illustrate a trend to offload attachable peripheral modules to the phone itself, as the cost and convenience of a standalone device overshadow a set of multiple accessories [31]. Despite many reported smartphone-based PoC sensors focusing on the device's camera to assess colorimetric/spectroscopic data, these approaches show as drawbacks the challenge of obtaining the same signals regardless of the diversity of cameras' specifications, quality, users' experience and environment light influence.

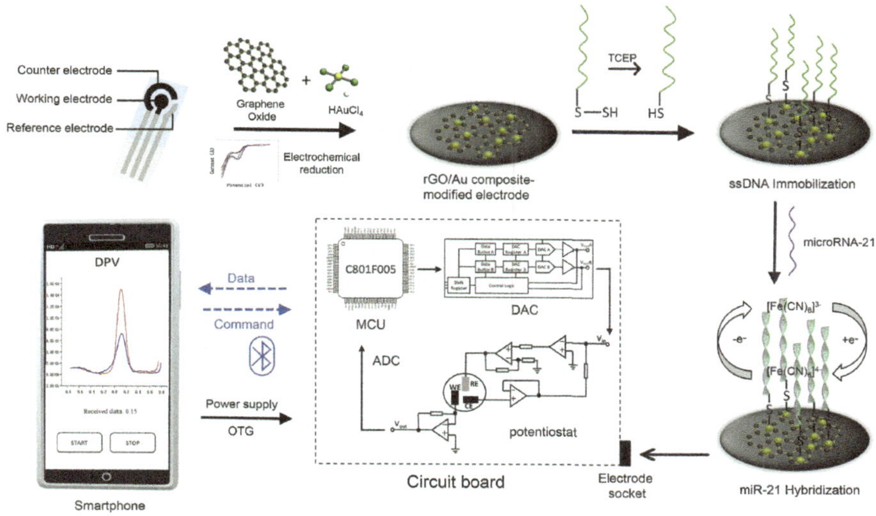

Fig. 8 Schematic of a smartphone-integrated PoC biosensor for microRNAs, the SPCE is functionalised with ssDNA via thiol linking; hybridisation with the complementary strand is detected by the changes in voltametric profile, using ferrocyanide as a redox reporter probe. A small potentiostat circuit is provided, powered by the phones' USB, and transmits data through Bluetooth. Reproduced from [132] with permission of the publisher

Electrochemical PoC biosensors are receiving further recognition as they eliminate these issues: They can easily be manufactured integrated into the phone's circuitry as onboard strip tests' interface. They can also benefit from phones' USB or audio jack port, and with the electronic meters and processing unit in wireless contact to the sensing electrode—via Bluetooth/4G/Wi-Fi/NFC, as shown in Fig. 8 [30]. Some selected articles using various electrochemical methods and operating via distinct interfaces are shown in Table 3; in most of them, a small printed circuit board (PCB) with embedded low-power processor unit coupled to a test-strip input does the electric measurement, a miniaturisation trend that nowadays is regarded as low cost.

5 Conclusions and Future Perspectives

Health care is an ongoing area of constant and fruitful findings, where growing demands for monitoring biological molecular markers is a permanent goal.

This chapter discussed the rich evolution of the biosensing area, in which multiple critical advances were made in the past decades to enable the development of small, portable, accurate and integrated biosensing devices. Most of these are still at a proof-of-concept stage; however, some are already available to monitor metabolites

Table 3 Selected examples of smartphone-based electrochemical PoC biosensors

Electrochemical method	Smartphone interface	Composition	Analyte/matrix	Obs	References
Cyclic voltammetry (CV)	Wi-Fi	PI/Graphene-ink/Cu^{2+}	Dopamine in sweat	Flexible sensor	[127]
	Bluetooth	SPCE/PB/CS/antigen + HRP-labelled antibody	Microcystin toxin in water	Small PCB, operational with μ-potentiostat	[128]
Differential pulse voltammetry (DPV)	Bluetooth	RGO/Au composite/ssDNA	miRNA in saliva	Small PCB powered by phone's USB port	[105]
	USB	Au@Fe_3O_4NP/ssDNA/thiol + calixarene/graphene/TB/AuNP/ssDNA	SARS-COV-2's ORF1ab gene in multiple biologic fluids	Sandwich-type assay (capture + label probes), excellent performance compared to RT-PCR	[129]
Chronoamperometry	Bluetooth	CNTs/BSA/FAD/TPP/GLA/PyOD	Phosphates in saliva	Screen-printed electrodes, read by portable μ-potentiostat	[130]
	USB	Chip + alginic acid/Ca^{2+} hydrogel + AOx + TMB + HRP	Ethanol in blood	Operational with μ-potentiostat, distinguishes legal/illegal levels for driving	[131]
	Test-strip connector in the phone	PET platform with GOx/ferrocyanide + β-HD/NAD^+	Glucose and blood ketones in whole-blood	Authors fabricated an electrochemical module integrated to phone's PCB; two tests in the same strip	[132]
		PET platform with ChOx/ferricyanide	Total cholesterol in blood	Smartphones' PCB adapted to include a test-strip port	[64]

(continued)

Table 3 (continued)

Electrochemical method	Smartphone interface	Composition	Analyte/matrix	Obs	References
	NFC	Au-coated SPCE/poly(β-CD)/antibody/BSA	Hepatitis B in blood serum	The NFC-potentiostat platform measures current and export data to the smartphone	[133]
Potentiometry	USB	SPCE/starch/ferricyanide	α-amylase in saliva	The small PCB attaches to the smartphones' USB port	[134]
EIS	Bluetooth	Interdigital Au electrodes/PEG-thiol/peptides + BSA	Thrombin in aqueous solution	Handheld test-strip analyser and data transmitter	[135]
		Si/PDMS/Au microfluidic platform	Blood cells counting in whole blood	Arduino PCB transmits data to be analysed by an Android App	[136]

PI polyimide, *PB* Prussian blue, *CS* chitosan, *HRP* horseradish peroxidase, *TB* toluidine blue, *TPP* thiamine pyrophosphate, *GLA* glutaraldehyde, *PyOD* pyruvate oxidase, *AOx* alcohol dehydrogenase, *TMB* tetramethylbenzidine, *β-CD* cyclodextrin, *ChOx* cholesterol oxidase, *β-HD* beta-hydroxybutyrate dehydrogenase and even neglected diseases in developing countries, although commercial success is still mostly reserved to glucometers.

When it comes to point-of-care, researchers are certainly feeling pressure to elaborate a larger variety of electrochemical biosensors while dealing with the challenges of stabilising the biological components for long-term storage, fine-tuning strategies to increase analytical performance, especially towards interferents, and preventing non-specific adsorption in biologic fluids. Some strategies are recently being explored to a greater extent, such as aptamers usage as stable biorecognition elements, hydrogels to provide a versatile and biomolecule-friendly environment and high-conductivity antifouling layers to prevent non-specific interaction to proteins and interferents, therefore increasing the lifetime and long-term sensitivity of the sensors. Future evolution in multiplex capability is also expected, so that a single skin patch should be able to monitor a variety of electrolytes and biomarkers, without the still existing problems of skin irritations.

In parallel, there is a strong demand for miniaturisation and integration to portable devices such as smartphones. This requires the engineering of not only sensing platforms or test-strips but also compatibility to personal glucometers, development of small handheld μ-potentiostat platforms or smartphone interfacing. Results are expected to be sent wirelessly to the phone, analysed by suitable software and stored online. Internet of things increase the pressure to interconnection, and one can expect PoC sensors operating with cloud computing, with the convenience of health data transmitted in real-time to physicians and hospitals, at the cost of the less mentioned issue of privacy and data sharing/leakage worries. Regulatory compliance should be another aspect of careful attention as multiple agencies and commissions require distinct data, stringent reports and granted certifications for biologic tests implementation, despite the track record of possible waivers.

PoC technology development should be enough to offer reliable and user-friendly—no sample pre-treatments required—biosensors to outweigh the also evolving trend of decentralised or "near point-of-care" laboratories, integrated into online information systems.

In low-income countries, the effects of PoC technology for improving life quality is depicted in studies, and the penetration of these tests through less accessible population groups will keep the advances to Global Health. The same promising characteristics of PoC devices can be expected to drive interest in their incorporation into military and defence affairs toolboxes, as well as to enforce environmental control and *on-site* veterinary screening of animal health.

Acknowledgements RNPC is grateful to FAPESP (process 2021/05665-7) for the funding.

References

1. Nagel B, Dellweg H, Gierasch LM (1992) Glossary for chemists of terms used in biotechnology (IUPAC Recommendations 1992). Pure Appl Chem 64:143–168

2. da Silva ETSG et al (2017) Electrochemical biosensors in point-of-care devices: recent advances and future trends. ChemElectroChem 4:778–794
3. Rengaraj S, Cruz-Izquierdo Á, Scott JL, Di Lorenzo M (2018) Impedimetric paper-based biosensor for the detection of bacterial contamination in water. Sensors Actuators B Chem 265:50–58
4. Choi J, Yong K, Choi J, Cowie A (2019) Emerging point-of-care technologies for food safety analysis. Sensors 19:817
5. Soper SA et al (2006) Point-of-care biosensor systems for cancer diagnostics/prognostics. Biosens Bioelectron 21:1932–1942
6. Jang A, Zou Z, Lee KK, Ahn CH, Bishop PL (2010) Potentiometric and voltammetric polymer lab chip sensors for determination of nitrate, pH and Cd(II) in water. Talanta 83:1–8
7. Su F et al (2017) Two-dimensional zirconium-based metal-organic framework nanosheet composites embedded with Au nanoclusters: a highly sensitive electrochemical aptasensor toward detecting cocaine. ACS Sensors 2:998–1005
8. Haleem A, Javaid M, Singh RP, Suman R, Rab S (2021) Biosensors applications in medical field: A brief review. Sensors Int 2:100100
9. Wang Y, Xu H, Zhang J, Li G (2008) Electrochemical sensors for clinic analysis. Sensors 8:2043–2081
10. Huang Y, Xu J, Liu J, Wang X, Chen B (2017) Disease-related detection with electrochemical biosensors: a review. Sensors 17:2375
11. Yager P, Domingo GJ, Gerdes J (2008) Point-of-care diagnostics for global health. Annu Rev Biomed Eng 10:107–144
12. Jeevanandam J, Barhoum A, Chan YS, Dufresne A, Danquah MK (2018) Review on nanoparticles and nanostructured materials: history, sources, toxicity and regulations. Beilstein J Nanotechnol 9:1050–1074
13. Yang W et al (2010) Carbon nanomaterials in biosensors: should you use nanotubes or graphene? Angew Chem Int Ed Engl 49:2114–2138
14. Shao Y et al (2010) Graphene based electrochemical sensors and biosensors: a review. Electroanalysis 22:1027–1036
15. Zhao H et al (2020) Nanofabrication approaches for functional three-dimensional architectures. Nano Today 30:100825
16. Zhang A, Bai H, Li L (2015) Breath figure: a nature-inspired preparation method for ordered porous films. Chem Rev 115:9801–9868
17. Biswas A et al (2012) Advances in top–down and bottom–up surface nanofabrication: Techniques, applications & future prospects. Adv Colloid Interface Sci 170:2–27
18. Welch NG, Scoble JA, Muir BW, Pigram PJ (2017) Orientation and characterization of immobilized antibodies for improved immunoassays (Review). Biointerphases 12:02D301
19. Liu Y, Yu J (2016) Oriented immobilization of proteins on solid supports for use in biosensors and biochips: a review. Microchim Acta 183:1–19
20. Shen M, Rusling JF, Dixit CK (2017) Site-selective orientated immobilization of antibodies and conjugates for immunodiagnostics development. Methods 116:95–111
21. Bilal M et al (2018) State-of-the-art protein engineering approaches using biological macromolecules: a review from immobilization to implementation view point. Int J Biol Macromol 108:893–901
22. Rashid JIA, Yusof NA (2017) The strategies of DNA immobilization and hybridization detection mechanism in the construction of electrochemical DNA sensor: a review. Sens. Bio-Sensing Res 16:19–31
23. Rasheed PA, Sandhyarani N (2017) Carbon nanostructures as immobilization platform for DNA: a review on current progress in electrochemical DNA sensors. Biosens Bioelectron 97:226–237
24. Oberhaus FV, Frense D, Beckmann D (2020) Immobilization techniques for aptamers on gold electrodes for the electrochemical detection of proteins: a review. Biosensors 10:45
25. Kudłak B, Wieczerzak M (2020) Aptamer based tools for environmental and therapeutic monitoring: a review of developments, applications, future perspectives. Crit Rev Environ Sci Technol 50:816–867

26. Zhang W, Wang R, Luo F, Wang P, Lin Z (2020) Miniaturized electrochemical sensors and their point-of-care applications. Chin Chem Lett 31:589–600
27. Liu D et al (2020) Trends in miniaturized biosensors for point-of-care testing. TrAC Trends Anal Chem 122:115701
28. Cirino NM, Musser KA, Egan C (2004) Multiplex diagnostic platforms for detection of biothreat agents. Expert Rev Mol Diagn 4:841–857
29. Liu J, Geng Z, Fan Z, Liu J, Chen H (2019) Point-of-care testing based on smartphone: the current state-of-the-art (2017–2018). Biosens Bioelectron 132:17–37
30. Gautam S, Batule BS, Kim HY, Park KS, Park HG (2017) Smartphone-based portable wireless optical system for the detection of target analytes. Biotechnol J 12:1600581
31. Sun AC, Hall DA (2019) Point-of-care smartphone-based electrochemical biosensing. Electroanalysis 31:2–16
32. Fan R, Andrew TL (2020) Perspective—challenges in developing wearable electrochemical sensors for longitudinal health monitoring. J Electrochem Soc 167:037542
33. Arab Hassani F et al (2020) Smart materials for smart healthcare– moving from sensors and actuators to self-sustained nanoenergy nanosystems. Smart Mater Med 1:92–124
34. Sharma A, Badea M, Tiwari S, Marty JL (2021) Wearable biosensors: an alternative and practical approach in healthcare and disease monitoring. Molecules 26:748
35. Katseli V, Economou A, Kokkinos C (2021) Smartphone-addressable 3D-printed electrochemical ring for nonenzymatic self-monitoring of glucose in human sweat. Anal Chem 93:3331–3336
36. Mohan AMV, Rajendran V, Mishra RK, Jayaraman M (2020) Recent advances and perspectives in sweat based wearable electrochemical sensors. TrAC Trends Anal Chem 131:116024
37. Kim J, Cha E, Park JU (2020) Recent advances in smart contact lenses. Adv Mater Technol 5:1–17
38. Dochez AR, Avery OT (1917) The elaboration of specific soluble substance by pneumococcus during growth. J Exp Med 26:477–493
39. Berson SA, Yalow RS (1959) Recent studies on insulin-binding antibodies. Ann N Y Acad Sci 82:338–344
40. Clark LC, Lyons C (1962) Electrode systems for continuous monitoring in cardiovascular surgery. Ann N Y Acad Sci 102:29–45
41. Guilbault GG, Montalvo JG (1970) Enzyme electrode for the substrate urea. J Am Chem Soc 92:2533–2538
42. Schläpfer P, Mindt W, Racine PH (1974) Electrochemical measurement of glucose using various electron acceptors. Clin Chim Acta 57:283–289
43. Varfolomeev SD, Berezin LV (1978) Enzymes as catalysts of electrochemical reactions. J Mol Catal 4:387–399
44. Bartlett PN, Al-Lolage FA (2017) There is no evidence to support literature claims of direct electron transfer (DET) for native glucose oxidase (GOx) at carbon nanotubes or graphene. J Electroanal Chem 1. https://doi.org/10.1016/j.jelechem.2017.06.021
45. Wilson GS (2016) Native glucose oxidase does not undergo direct electron transfer. Biosens Bioelectron 82:vii–viii
46. Gonçales VR et al (2016) Three-dimensional graphene/carbon nanotubes hybrid composites for exploring interaction between glucose oxidase and carbon based electrodes. J Electroanal Chem 775:235–242
47. Engvall E, Perlmann P (1971) Enzyme-linked immunosorbent assay (ELISA) quantitative assay of immunoglobulin G. Immunochemistry 8:871–874
48. Liedberg B, Nylander C, Lunström I (1983) Surface plasmon resonance for gas detection and biosensing. Sensors Actuators 4:299–304
49. Shichiri M, Yamasaki Y, Kawamori R, Hakui N, Abe H (1982) Wearable artificial endocrine pancreas with needle-type glucose sensor. Lancet 320:1129–1131
50. Cass AEG et al (1984) Ferrocene-mediated enzyme electrode for amperometric determination of glucose. Anal Chem 56:667–671

51. Matthews DR et al (1987) Pen-sized digital 30-second blood glucose meter. Lancet 329:778–779
52. Wang J (2001) Glucose biosensors: 40 years of advances and challenges. Electroanalysis 13:983–988
53. Yoo E-H, Lee S-Y (2010) Glucose biosensors: an overview of use in clinical practice. Sensors 10:4558–4576
54. Campbell RL, Wagner DB, O'Connell JP (1984) Solid phase assay with visual readout
55. Kost GJ et al (2000) Multicenter study of whole-blood creatinine, total carbon dioxide content, and chemistry profiling for laboratory and point-of-care testing in critical care in the United States. Crit Care Med 28:2379–2389
56. Louie RF, Tang Z, Shelby DG, Kost GJ (2000) Point-of-care testing: millennium technology for critical care. Lab Med 31:402–408
57. Gilbert HC, Szokol JW (2004) Point of care technologies. Int Anesthesiol Clin 42:73–94
58. de Vasconcelos EA et al (2009) Potential of a simplified measurement scheme and device structure for a low cost label-free point-of-care capacitive biosensor. Biosens Bioelectron 25:870–876
59. Ahn CH et al (2004) Disposable smart lab on a chip for point-of-care clinical diagnostics. Proc IEEE 92:154–173
60. Zheng G, Patolsky F, Cui Y, Wang WU, Lieber CM (2005) Multiplexed electrical detection of cancer markers with nanowire sensor arrays. Nat Biotechnol 23:1294–1301
61. Gates BD et al (2005) New approaches to nanofabrication: molding, printing, and other techniques. Chem Rev 105:1171–1196
62. Gao W et al (2016) Fully integrated wearable sensor arrays for multiplexed in situ perspiration analysis. Nature 529:509–514
63. Vasudev A, Kaushik A, Tomizawa Y, Norena N, Bhansali S (2013) An LTCC-based microfluidic system for label-free, electrochemical detection of cortisol. Sensors Actuators B Chem 182:139–146
64. Fu Y, Guo J (2018) Blood cholesterol monitoring with smartphone as miniaturized electrochemical analyzer for cardiovascular disease prevention. IEEE Trans Biomed Circuits Syst 12:784–790
65. Mannoor MS et al (2012) Graphene-based wireless bacteria detection on tooth enamel. Nat Commun 3:763
66. Yoon S et al (2021) Multifunctional hybrid skin patch for wearable smart healthcare applications. Biosens Bioelectron 113685. https://doi.org/10.1016/j.bios.2021.113685
67. Sheng X et al (2021) Point-of-care monitoring of colitis using intestinal alkaline phosphatase in inflammatory bowel disease. ACS Sensors 6:698–702
68. Xu J et al (2012) Sensitive point-of-care monitoring of HIV related DNA sequences with a personal glucometer. Chem Commun 48:10733
69. Narang J et al (2018) Portable bioactive paper based genosensor incorporated with Zn-Ag nanoblooms for herpes detection at the point-of-care. Int J Biol Macromol 107:2559–2565
70. Tseng CC et al (2021) Rapid electrochemical-biosensor microchip platform for determination of microalbuminuria in CKD patients. Anal Chim Acta 1146:70–76
71. Kamakoti V, Kinnamon D, Choi K H, Jagannath B, Prasad S (2018) Fully electronic urine dipstick probe for combinatorial detection of inflammatory biomarkers. Futur Sci OA 4:FSO301
72. Chen C-H, Lin MS (2012) A novel structural specific creatinine sensing scheme for the determination of the urine creatinine. Biosens Bioelectron 31:90–94
73. Sánchez-Tirado E, Martínez-García G, González-Cortés A, Yáñez-Sedeño P, Pingarrón JM (2017) Electrochemical immunosensor for sensitive determination of transforming growth factor (TGF)—β1 in urine. Biosens Bioelectron 88:9–14
74. Pan Y et al (2010) Electrochemical immunosensor detection of urinary lactoferrin in clinical samples for urinary tract infection diagnosis. Biosens Bioelectron 26:649–654
75. Lam B, Fang Z, Sargent EH, Kelley SO (2012) Polymerase chain reaction-free, sample-to-answer bacterial detection in 30 minutes with integrated cell lysis. Anal Chem 84:21–25

76. Ogden NE, Kurnik M, Parolo C, Plaxco KW (2019) An electrochemical scaffold sensor for rapid syphilis diagnosis. Analyst 144:5277–5283
77. Uludag Y et al (2016) An integrated lab-on-a-chip-based electrochemical biosensor for rapid and sensitive detection of cancer biomarkers. Anal Bioanal Chem 408:7775–7783
78. UNODC (2016) World drug report 2021. in 11–11. https://doi.org/10.18356/a6fa3135-en
79. Hassan SA, ElDin NB, Zaazaa HE, Moustafa AA, Mahmoud AM (2020) Point-of-care diagnostics for drugs of abuse in biological fluids: application of a microfabricated disposable copper potentiometric sensor. Microchim Acta 187:491
80. Tanak AS, Muthukumar S, Hashim IA, Prasad S (2020) Establish pre-clinical diagnostic efficacy for parathyroid hormone as a point-of-surgery-testing-device (POST). Sci Rep 10:1–9
81. Gattani A, Singh SV, Agrawal A, Khan MH, Singh P (2019) Recent progress in electrochemical biosensors as point of care diagnostics in livestock health. Anal Biochem 579:25–34
82. Turner AP, Magan N (2004) Electronic noses and disease diagnostics. Nat Rev Microbiol 2(2):161–166
83. Gabig-Ciminska M, Andresen H, Albers J, Hintsche R, Enfors SO (2004) Identification of pathogenic microbial cells and spores by electrochemical detection on a biochip. Microb Cell Fact 3:1–11
84. MacLeod JA et al (2016) Fast, sensitive point of care electrochemical molecular system for point mutation and select agent detection. Lab Chip 16:2513–2520
85. Batista Deroco P, Wachholz Junior D, Tatsuo Kubota L (2021) Recent advances in point-of-care biosensors for the diagnosis of neglected tropical diseases. Sensors Actuators B Chem 130821. https://doi.org/10.1016/j.snb.2021.130821
86. Garrote BL, Santos A, Bueno PR (2020) Label-free capacitive assaying of biomarkers for molecular diagnostics. Nat Protoc 15:3879–3893
87. Lenyk B et al (2020) Dual-transducer malaria aptasensor combining electrochemical impedance and surface plasmon polariton detection on gold nanohole arrays. ChemElectroChem 7:4594–4600
88. Rosa BMG, Yang GZ (2020) Portable impedance analyzer as a rapid screening tool for malaria: an experimental study with culture and blood infected samples by early forms of plasmodium falciparum. IEEE Trans Biomed Eng 67:3531–3541
89. Santos GS et al (2019) Label-free nanostructured biosensor for Schistosoma mansoni detection in complex biological fluids. Talanta 204:395–401
90. Singhal C, Khanuja M, Chaudhary N, Pundir CS, Narang J (2018) Detection of chikungunya virus DNA using two-dimensional MoS2 nanosheets based disposable biosensor. Sci Rep 8:7734
91. Nazari-Vanani R, Heli H, Sattarahmady N (2020) An impedimetric genosensor for Leishmania infantum based on electrodeposited cadmium sulfide nanosheets. Talanta 217:121080
92. Cajigas S, Alzate D, Orozco J (2020) Gold nanoparticle/DNA-based nanobioconjugate for electrochemical detection of Zika virus. Microchim Acta 187:594
93. Chen Y et al (2017) Field-effect transistor biosensor for rapid detection of Ebola antigen. Sci Rep 7:10974
94. Jin X et al (2019) A field effect transistor modified with reduced graphene oxide for immunodetection of Ebola virus. Microchim Acta 186:223
95. Mattioli IA, Hassan A, Oliveira ON, Crespilho FN (2020) On the challenges for the diagnosis of SARS-CoV-2 based on a review of current methodologies. ACS Sensors 5:3655–3677
96. Mattioli IA, Crespilho FN (2020) Problems of interpreting diagnostic tests for SARS-CoV-2: analytical chemistry concerns. An Acad Bras Cienc 92
97. Torres MDT, de Araujo WR, de Lima LF, Ferreira AL, de la Fuente-Nunez C (2021) Low-cost biosensor for rapid detection of SARS-CoV-2 at the point of care. Matter 4:2403–2416
98. Narang J et al (2017) Point of care with micro fluidic paper based device integrated with nano zeolite–graphene oxide nanoflakes for electrochemical sensing of ketamine. Biosens Bioelectron 88:249–257

99. da Costa Oliveira T, Santana MHP, Banks CE, Munoz RAA, Richter EM (2018) Electrochemical portable method for on-site screening of scopolamine in beverage and urine samples. Electroanalysis elan.201800707. https://doi.org/10.1002/elan.201800707
100. Akyazi T, Basabe-Desmonts L, Benito-Lopez F (2018) Review on microfluidic paper-based analytical devices towards commercialisation. Anal Chim Acta 1001:1–17
101. Dungchai W, Chailapakul O, Henry CS (2009) Electrochemical detection for paper-based microfluidics. Anal Chem 81:5821–5826
102. Cunningham JC, Brenes NJ, Crooks RM (2014) Paper electrochemical device for detection of DNA and thrombin by target-induced conformational switching. Anal Chem 86:6166–6170
103. Silva ETSG, Santhiago M, Barragan JTC, Kubota LT (2014) Construction of a new versatile point-of-care testing device with electrochemical detection employing paper as a microfluidic platform. Anal Methods 6:6133–6136
104. Gao W (2015) The chemistry of graphene oxide. Graphene Oxide Reduct Recipes Spectrosc Appl 61–95. https://doi.org/10.1007/978-3-319-15500-5_3
105. Shin Low S et al (2020) Smartphone-based portable electrochemical biosensing system for detection of circulating microRNA-21 in saliva as a proof-of-concept. Sensors Actuators, B Chem 308:127718
106. Wang C et al (2016) A label-free and portable graphene FET aptasensor for children blood lead detection. Sci Rep 6:21711
107. Lu J, Ge S, Ge L, Yan M, Yu J (2012) Electrochemical DNA sensor based on three-dimensional folding paper device for specific and sensitive point-of-care testing. Electrochim Acta 80:334–341
108. Moazzam P et al (2021) Ultrasensitive detection of programmed death-ligand 1 (PD-L1) in whole blood using dispersible electrodes. Chem Commun 57:2559–2562
109. Henry AC et al (2000) Surface modification of poly(methyl methacrylate) used in the fabrication of microanalytical devices. Anal Chem 72:5331–5337
110. Rossier J, Reymond F, Michel PE (2002) Polymer microfluidic chips for electrochemical and biochemical analyses. Electrophoresis 23:858–867
111. da Silva AC, Córdoba de Torresi SI (2019) Advances in conducting, biodegradable and biocompatible copolymers for biomedical applications. Front Mater 6
112. Sonawane A, Manickam P, Bhansali S (2017) Stability of enzymatic biosensors for wearable applications. IEEE Rev Biomed Eng 10:174–186
113. Sharma S, Byrne H, O'Kennedy RJ (2016) Antibodies and antibody-derived analytical biosensors. Essays Biochem 60:9–18
114. Sabaté del Río J, Henry OYF, Jolly P, Ingber DE (2019) An antifouling coating that enables affinity-based electrochemical biosensing in complex biological fluids. Nat Nanotechnol 14:1143–1149
115. Odenthal KJ, Gooding JJ (2007) An introduction to electrochemical DNA biosensors. Analyst 132:603
116. Mearns FJ, Wong ELS, Short K, Hibbert DB, Gooding JJ (2006) DNA biosensor concepts based on a change in the DNA persistence length upon hybridization. Electroanalysis 18:1971–1981
117. Farzin L, Shamsipur M, Samandari L, Sheibani S (2018) Advances in the design of nanomaterial-based electrochemical affinity and enzymatic biosensors for metabolic biomarkers: a review. Microchim Acta 185:276
118. Sassolas A, Blum LJ, Leca-Bouvier BD (2009) Electrochemical aptasensors. Electroanalysis 21:1237–1250
119. Couto RAS, Lima JLFC, Quinaz MB (2016) Recent developments, characteristics and potential applications of screen-printed electrodes in pharmaceutical and biological analysis. Talanta 146:801–814
120. de Oliveira TR, Fonseca WT, de Oliveira Setti G, Faria RC (2019) Fast and flexible strategy to produce electrochemical paper-based analytical devices using a craft cutter printer to create wax barrier and screen-printed electrodes. Talanta 195:480–489

121. Middya S, Kaminski Schierle GS, Malliaras GG, Curto VF (2021) Lithography and electrodes. In: Chemical solution synthesis for materials design and thin film device applications. Elsevier, pp. 277–307. https://doi.org/10.1016/B978-0-12-819718-9.00005-4
122. Norrman K, Ghanbari-Siahkali A, Larsen NB (2005) Studies of spin-coated polymer films. Annu Rep Prog Chem Sect C 101:174–201
123. Cheng R et al (2020) Porous graphene oxide films prepared via the breath-figure method: a simple strategy for switching access of redox species to an electrode surface. ACS Appl Mater Interfaces 12:55181–55188
124. Urban GA, Weiss T (2009) Hydrogels for biosensors, pp. 197–220. https://doi.org/10.1007/978-3-540-75645-3_6
125. Bahia K, Delaporte A (2020) The state of mobile internet connectivity 2020. GSMA Reports 61
126. Bahia K, Delaporte A (2021) The state of mobile internet connectivity 2021. GSMA Reports 61
127. Muralidharan R, Chandrashekhar V, Butler D, Ebrahimi A (2020) A smartphone-interfaced, flexible electrochemical biosensor based on graphene ink for selective detection of dopamine. IEEE Sens J 20:13204–13211
128. Guan T et al (2019) Point-of-need detection of microcystin-LR using a smartphone-controlled electrochemical analyzer. Sensors Actuators B Chem 294:132–140
129. Zhao H et al (2021) Ultrasensitive supersandwich-type electrochemical sensor for SARS-CoV-2 from the infected COVID-19 patients using a smartphone. Sensors Actuators B Chem 327:128899
130. Bai Y et al (2021) An inkjet-printed smartphone-supported electrochemical biosensor system for reagentless point-of-care analyte detection. Sensors Actuators B Chem 346:130447
131. Aymerich J et al (2018) Cost-effective smartphone-based reconfigurable electrochemical instrument for alcohol determination in whole blood samples. Biosens Bioelectron 117:736–742
132. Guo J, Huang X, Ma X (2018) Clinical identification of diabetic ketosis/diabetic ketoacidosis acid by electrochemical dual channel test strip with medical smartphone. Sensors Actuators B Chem 275:446–450
133. Teengam P et al (2021) NFC-enabling smartphone-based portable amperometric immunosensor for hepatitis B virus detection. Sensors Actuators B Chem 326:128825
134. Zhang L, Yang W, Yang Y, Liu H, Gu Z (2015) Smartphone-based point-of-care testing of salivary α-amylase for personal psychological measurement. Analyst 140:7399–7406
135. Zhang D et al (2016) Protein detecting with smartphone-controlled electrochemical impedance spectroscopy for point-of-care applications. Sensors Actuators B Chem 222:994–1002
136. Talukder N et al (2017) A portable battery powered microfluidic impedance cytometer with smartphone readout: towards personal health monitoring. Biomed Microdevices 19:36

The manufacturer's authorised representative in the EU is Springer Nature Customer Service Centre GmbH, Europaplatz 3, 69115 Heidelberg, Germany. If you have any concerns regarding our products, please contact ProductSafety@springernature.com

Printed and bound by CPI Group (UK) Ltd, Croydon, CR0 4YY
30/09/2026
02225329-0002